Springer Series in Materials Science

Volume 202

For further volumes:
http://www.springer.com/series/856

The Springer Series in Materials Science covers the complete spectrum of materials physics, including fundamental principles, physical properties, materials theory and design. Recognizing the increasing importance of materials science in future device technologies, the book titles in this series reflect the state-of-the-art in understanding and controlling the structure and properties of all important classes of materials.

C. M. Sunta

Unraveling Thermoluminescence

Springer

C. M. Sunta
Radiation Protection
Formerly from Bhabha Atomic
Research Center and Atomic Energy
Regulatory Board
Government of India
Mumbai, Maharashtra
India

ISSN 0933-033X ISSN 2196-2812 (electronic)
ISBN 978-81-322-1939-2 ISBN 978-81-322-1940-8 (eBook)
DOI 10.1007/978-81-322-1940-8
Springer New Delhi Heidelberg New York Dordrecht London

Library of Congress Control Number: 2014941507

Printed on acid-free paper

Springer is part of Springer Science+Business Media (www.springer.com)

अपूर्वः कोपि कोशोऽयं
विधते तव भारती ।
व्ययतो वृध्दिम् आयाति
क्षयम् आयाति
संचयात ।।

O, Saraswati ! Goddess of wisdom and learning,
How unique your treasure house of knowledge is.
It reduces if you hoard it, and
becomes richer when you share it !

Foreword

During the past few decades, thermoluminescence (TL) and related techniques have provided substantive inputs in the field of radiation dosimetry. Today most of the radiation dosimetry applications, be it the environmental, personnel dosimetry, space radiation, geo- and archeo-chronology to the estimation of dose delivered to a patient under radiation therapy are based on TLDs and OSL detectors, with TL offering a distinct advantage of its ability to isolate signals/traps that appear at different temperatures/energies. The traffic of charges between the traps has always been a deeply contested issue and there has been no consensus on any mechanism that can adequately describe the underlying kinetic processes and a physical description of these.

This book by Dr. Sunta presents a state-of-the-art account of the current understanding of the theoretical aspects of TL and presents a lucid account of the basics and includes several new aspects of interpretations that are not found in text books so far. The book assumes a basic knowledge of Physics and Mathematics at an undergraduate level, but then takes the reader to the results at the cutting edge of understanding of the subject in an easy to understand manner. The book begins by presenting an introduction to the basic TL phenomena and the point defects in inorganic solids that may serve as traps and recombination centers. It then describes various models that describe kinetics and shows how inadequate the understanding has been in the past. It then takes the next step to explain methods for the kinetic analysis of TL glow peaks and not only discusses the conventional methods but also presents some new possibilities. It then deals with the concept of quasi-equilibrium and discusses various related issues and presents several dimensions of quasi-equilibrium approximation. As the next logical step it then discusses the growth of TL intensity with radiation dose and here highlights the role of thermally disconnected traps—a concept used in the past, but never given its due. The author derives the supralinearity factors and pre-dose sensitization factors from first principles, and presents a new factor called the sensitization factor to provide a unified explanation to two seemingly diverse aspects of supralinearity and pre-dose effect. The next chapter discusses Photo transferred TL in all possible scenarios and with case studies including its various applications.

I found that this book presents the distilled wisdom of the author Dr. C. M. Sunta, who has a nearly five-decade long experience as a 24 × 7 TL practitioner and as a pioneer in the conceptual development of the subject. I have myself treated Dr. Sunta as one of my teachers in Luminescence dosimetry and I feel happy that such a book will soon be available to students, teachers, and practitioners and will serve as a useful handbook on concepts and formalism. I feel that this will be a must-read for any one venturing into TL and its application and I am sure this will enrich their science manifold. I compliment and thank Dr. Sunta for his assiduous efforts to present his experience to the younger fraternity.

Ahmedabad, India A. K. Singhvi

Preface

One may perhaps question the need for another book on this subject when so many of them have been written in the past.

The title of the book—"Unravelling Thermoluminescence" in some way provides an answer to this query. The dictionary defines "unravel" as an attempt to "make plain or clear; solve or explain something complicated or puzzling." That is what this book seeks to do.

The first two chapters are written assuming that the reader is a fresh undergraduate who is beginning to take up work in the field of thermoluminescence (TL). The subject of TL, being one among the variety of luminescence phenomena in nature, it is thought fit to acquaint the student with these various phenomena.

Chapter 1 includes the historical development of the subject of TL and also gives a brief overview of the luminescence phenomena.

Chapter 2 summarizes well-known types of point defects in the inorganic solids. These serve as simple examples of defect centres, the types of which may act as traps and recombination centers. However, the actual structure of the defect centers in different TL phosphors is not simple to find out, neither is it necessary for the application of TL in areas like dosimetry and archeological and geological dating. X–rays and nuclear radiations being the principal sources which induce the TL, their interaction with matter is summarized in this chapter. The interactions are discussed in the perspective of the TL induction by different types of radiations.

The subsequent chapters go on to address those points that are new or those that have not been included in the books published earlier. Such parts have been highlighted by showing them in italics. The stress has been more on explaining the concepts than reviewing the earlier works. In the process reference to earlier literature is not comprehensive.

Chapter 3 begins with the explanation of the thermal stimulation process and goes on to describe conventional models of TL. The characteristic properties of the glow curves of the Randall–Wilkins, the Garlick–Gibson, the general order (GO), and the mixed order (MO) kinetics models are summarized. *While dealing with the GO kinetics model it is shown that the pre-exponential factor s' and the kinetic order (KO) parameter b are not independent constants as assumed by the* *proponents of the model, but that the s' value is dependent on the value of b and the total concentration N of the traps. When GO kinetics model is applied to the simple one trap one recombination center (OTOR) physical model, it is seen that the b value itself varies with n_0/N, which means it varies with dose given to the sample. The lacunae in GO kinetics model are discussed in Sects. 3.3.4 and 3.3.4.1. The MO kinetics model too has drawbacks. A major drawback of MO model is that its KO parameter α is dependent on n_0, which means that in real samples it would depend on the dose given to the sample. Thus no unique value of α may be assigned to a given glow peak of a sample. The chapter then takes up the physical models. These include the multi-trap systems which are the simplified version of the generalized scheme consisting of a host of traps and an equally large number of recombination centers. These may plausibly be applicable to the real materials. It is shown by simulations how first order (FO) kinetics glow peaks are produced under a variety of parametric conditions. The results of these simulations are used to answer the question why the KO of the TL glow peaks of real materials is invariably seen to be of FO. The conclusion drawn is that there is an abundance of thermally disconnected deep traps which provide for the recombination rate to exceed the retrapping rate and thus produce FO kinetics. Further, at low trap occupancies, FO is produced even in retrapping dominant conditions. It is logical that defects would exist in a crystalline material until it approaches its melting point. Examples are given of some materials in which deep traps have been detected. Another way by which FO kinetics is produced is the local recombination of the thermally excited charge carriers. Examples are given from the published papers wherein local recombination is proved.*

Chapter 4 deals with the kinetics analysis of the TL glow peaks. Five methods of glow curve analysis are summarized including the merits and the demerits of each method and the precaution in their application. The parts which I may call worth mentioning as new in this chapter are as follows:

(1) *A stable peak temperature T_m irrespective of trap occupancy n_0/N (dose) is the property exclusive to the FO kinetics. In all non-FO cases the glow peak temperature changes with trap occupancy. This property may be used to differentiate the non-FO cases if any, from the FO ones before attempting the kinetics analysis of any experimental glow curve which requires kinetic order.* Application of peak shape method in analyzing the experimental glow curves may face uncertainty in the shape due to the presence of weak satellites close to the peak under study.

(2) The peak temperature of the simulated TL glow peaks is found to vary linearly with the activation energy E for a given s/β. Examples of real samples are given which also show that activation energy E of different glow peaks of the given sample are almost linearly related to the glow peak temperature. This implies that *the value of the frequency factor s for the different glow peaks of a given sample may be same*.

(3) *A method has been suggested to find the value of KO using the initial rise profiles. It is a new method which has been proposed for the first time. The method is general in the sense that it is applicable to all values of KO.*

(4) *If a GO kinetics glow curve is analyzed by using the isothermal decay method, it turns out that the resulting E and the KO values are dependent on the initial concentration n_0 of the active traps (dose). This result once again brings out the lacunae in the GO kinetics expression.*

(5) *Application of the peak fitting method to the simulated glow peaks of different physical models shows that the error in the found value of E and the minimum error in fitting vary with the best fitted value of b. The found value of E turns out to be error free only when the corresponding best fitted value of b is either 1 or 2. For cases in which the best fit value of b lies between 1 and 2, error in found E value is seen to be highest when the value of b is 1.5. At this point the error value also is seen to be highest. The highest error in found value of E at this point is about ±8 %. For best fit b values less than 1 or more than 2, error in found value of E are much higher and so are the minimized fitting error values.*

(6) An example taken from a published paper on post-irradiation annealed LiF: Mg, Ti sample shows that the obtained values of *E* as well as *KO* using the curve fitting method increase with the annealing time. When we examine the analysis procedure used by the authors of the paper we see that the minimum value of the error function (FOM) also has increased with the annealing time. This implies that *the minimization of the error function in curve fitting method does not ensure the accuracy in the obtained value of E if the best fit value of KO is found to be different from 1 or 2.*

Chapter 5 deals with the quasi-equilibrium (QE) problem in TL. To the knowledge of this author a large part of the material of this chapter has not appeared in earlier books in this field. *It is concluded in certain earlier publications that QE cannot be achieved under the condition in which retrapping rate is much faster than the recombination rate. In this chapter this conclusion is disproved and it is shown that it is the values of the system parameters, namely the concentration of the traps and the recombination centers and their cross sections, which determine whether or not the QE condition would be satisfied. In this connection the importance of the recombination lifetime is highlighted. Smaller the value of the recombination lifetime, better it is for the QE to be achieved.* Using the premise that under the QE condition the relaxation rate of the thermally excited carriers should be equal to the excitation rate, *the minimum value of the function $\Sigma_i N_i \sigma_i$ for the validity of the QE approximation is found, where N_i stands for the concentrations of the active traps and the deeper traps and σ_i stands for their respective cross sections. This is a new approach to find the range of parameter values to satisfy the QE condition.*

It is shown for the first time that the QE depends not only on the parameter values of the system but also on the heating rate. At a certain high heating rate, depending on the parameter values of the system, the QE begins to deteriorate, which reflects as change in the peak shape. It is suggested that the heating rate dependence of QE may be used to test whether or not QE approximation holds for an experimental glow curve. It is shown that the heating rate dependence of QE may be used also to find the range of the values of parameters (trap populations and the cross sections) at least approximately, for an experimental glow peak. As an illustration this test is applied to the glow peak No. 5 of LiF TLD phosphor.

Chapter 6 deals with the growth of TL intensity with radiation dose. After a brief review of earlier theories and earlier works on this subject, the discussion takes up a model which gives physical meaning to the so-called competitors during the TL read out heating. *As against the arbitrarily assumed competitors by the earlier workers, in the proposed model the competitors are the thermally disconnected deep traps (TDDT) which when empty act as competitors and when filled cause increase in the number of recombination centres.* When the active traps and the TDDTs get filled up during irradiation the reduction in competition and the increase in filled active trap population take place simultaneously. As a result the TL intensity growth becomes superlinear. The treatment of the model takes into count the irradiation and the heating stages together. *Using this model expressions are derived for the TL intensity growth curve, the supralinearity factor (SF), and the pre-dose sensitization factor (PDSF). Also derived is a new expression called the sensitization factor (S_nF). The factor S_nF which has been introduced by this author is used to give a unified explanation for the mechanisms which give rise to the SF and the PDSF. The computed profiles of SF, PDSF, and S_nF are compared with the experimentally obtained profiles of these factors for the case of the LiF:Mg, Ti TLD phosphor. These are found to match well with each other supporting the soundness of the model used.* Apart from providing a theoretical basis underlying the phenomena of supralinearity and pre-dose sensitization, the model dispels the doubt of some workers that the mechanism involved in these two phenomena may be altogether different from each other.

Chapter 7 deals with photo-transfer of TL (PTTL) intensity from one glow peak to another. This phenomenon is observed almost in all thermoluminescent materials, usually showing a transfer from higher temperature glow peak to those of lower temperatures. *In this chapter transfer is described from a glow peak to lower as well as to higher temperature glow peaks. This property has not been described in earlier books. This chapter also describes the effect of sample temperature on the transfer efficiency for incident light. This phenomenon is used to find the energy levels of the donor trap. Using the monochromatic incident light for photo-transfer, optical trap depth is found for the donor trap.* Studies describing the dependence of transfer efficiency on sample temperature as well as studies with monochromatic light exposure for photo-transfer have possibly not been given in earlier books of this field. *These studies are used to propose a model for photo-transfer in mineral CaF_2.* Finally this chapter deals with photo-bleaching, which is an important subject in sediment dating. It is shown that in a multi-peak glow curve photo-bleaching and photo-transfer run together. On prolonged light exposure, all glow peaks, donors, as well as accepters attain a low and nearly stable level. At this stage most of the traps are empty and correspondingly the population of the recombination centers also is negligibly small. *As a result the available option for the photo-excited electrons from any of the traps is simply to get retrapped into any of the vacant traps. In other words a to and fro transfer between the traps continues almost indefinitely. This leads to a nearly unbleacheable TL signal which is used as the zero level in the TL dating of the sediments.*

Mumbai

C. M. Sunta

Acknowledgments

I owe my deepest gratitude to my late teacher Dr. A. K. Ganguly who assigned to me the work of developing a thermoluminescence (TL) dosimetry system way back around 1965 for use in the nuclear industry in India.

A major part of this book, which supplements earlier books, is the result of the work I did at the University of Sao Paulo (USP), Brazil. During 1992–1994, my visit to USP was arranged by Emico Okuno. I am thankful to her and to Elizabeth Yoshimura for the work we did jointly at USP. This work appears in Chap. 6 of this book. The collaboration at USP was continued with Prof. Shigueo Watanabe. I would remain ever thankful to Prof. Watanabe, who in addition to giving academic support was extremely hospitable. He and his wife Dona Renate looked after every need and comfort for me and my wife during our visits to Brazil till mid 2004. I owe special thanks to J. F. D. Chubaci and his wife Rosinha who provided every help in and outside the institute in Brazil. I am thankful to my co-workers, J. F. D. Chubaci, Walter Ayta, Tom Piter, and late R. N. Kulkarni, who contributed immensely to the work we did at USP, part of which appears in Chaps. 3, 4 and 5 of this book.

Though this book was conceptualized and written after my retirement I am thankful to B. C. Bhatt, M. S. Kulkarni, Munish Kumar, Anuj Soni, and N. S. Rawat from BARC who have given substantial assistance in preparing the manuscript. It is no small favor to read a friend's manuscript. B. C. Bhatt and Vinita Puri have been gracious enough to read the manuscript and give valuable suggestions and inputs. However, I bear the responsibility for any mistakes or deficiencies either in substance or language that might have remained unattended. My deep apologies to all those whose names I might have missed to acknowledge.

My boundless thanks to my wife Kamla for her patience, encouragement, and sound advice during the course of writing this book. She stayed the course with me and has often had to realign some of her own priorities in the interests of this book. I thank my children, their spouses, and my grandchildren who have been unstinting in their support and love.

Mumbai C. M. Sunta

Contents

About the Author

Dr. C. M. Sunta was awarded the M.Sc. degree in Physics from the University of Agra before joining the Bhabha Atomic Research Center. He was awarded the Master of Science degree from University of Rochester, New York during his stint as a Health Physics trainee at Rochester and Brookhaven National Laboratory, New York. He was awarded the Ph.D. in Physics during his research career at Bhabha Atomic Research Centre (BARC), Mumbai. During 1968–1975, he worked on two major Research projects, one under World Health Organization (WHO) and the other under the International Atomic Energy Agency (IAEA) Vienna. The former dealt with exhaustive measurement of background radiations along the coastal areas of Kerala State of India which have radioactivity due to thorium bearing sand and its possible health effects. The IAEA project dealt with the development of radiation monitoring systems using the thermoluminescence (TL) phenomenon. Dr. Sunta continued his basic and applied research on TL simultaneously and heading the research group on this subject. From 1973 to 1975, he worked as an International Atomic Energy Agency's expert in Brazil, helping researchers there develop TL dosimetry systems and enhance applied research in this area. He has 80 peer reviewed papers in journals and about 110 conference presentations and research reports to his credit. In 1990, he was appointed as Director, Radiation Safety Division in the Atomic Energy Regulatory Board (AERB), Govt. of India, from where he superannuated in 1992. Subsequently, he accepted a research assignment in the department of Nuclear Physics at the University of Sao-Paulo Brazil and continued as a visiting Professor till 2004. He continues to pursue his research interest interacting with young researchers of BARC. He is a Fellow of the Luminescence Society of India and a Life Member of the Indian Nuclear Society, the Indian Association for Radiation Protection and the Radiation Biology Association of India.

About the Author

Chapter 1
Introduction: Thermoluminescence and Other Forms of Luminescence

Abstract Historical development of the subject of thermoluminescence (TL) is given since its discovery by Robert Boyle in 1663. Highlights of early period include the use of TL detectors in early rocket flights, which lead to the discovery X-rays in Solar spectrum. A brief overview of various types of luminescence phenomena such as photoluminescence, radioluminescence, chemiluminescence, bioluminescence, etc., also is given.

1.1 Thermoluminescence: Definition

The term "thermoluminescence" (TL) consists of two words: thermo, meaning heat and luminescence, meaning emission of light. These words may appear to mean that the emission of luminescence is caused by heat. This however is not totally true, since heat has only a secondary role in this case. When heat is the primary cause, the emission is called incandescence, for example, the light coming from a hot filament lamp or a burning charcoal. The primary agents for the induction of TL in a material are the ionizing radiations, namely the X-rays, or radiations from radioactive elements to which the sample is pre-exposed. In some materials, ultraviolet light also may excite TL. When the sample is subsequently warmed up, luminescent light is emitted. The act of warming up stimulates the release of the stored energy in the sample due to pre-exposure to the ionizing radiations. The release of this stored energy produces the luminescence. Therefore, some authors have appropriately used an expanded expression for this emission calling it *radiation induced thermally stimulated luminescence.* Some others have abbreviated it to *radiothermoluminescence* to signify that the TL is induced by the radiations. Some authors call it *thermally stimulated luminescence* (TSL). Presently however, most workers of this field have become accustomed to the abbreviated term TL. Incandescence appears in the visible wavelengths only when the temperature of the heated object goes above about 500 °C when a faint red glow begins to appear. As the temperature is raised further, the object becomes

C. M. Sunta, *Unraveling Thermoluminescence*,
Springer Series in Materials Science 202,
DOI: 10.1007/978-81-322-1940-8_1, © Springer India 2015

brighter and brighter. The temperature of the tungsten filament in a light bulb may go up to about 2,500 °C. In contrast, TL may appear at any temperature. It appears in the form of a transient glow, when the sample is heated for the first time after exposure to the ionizing radiations. Repeat heating does not produce any emission, unless the sample is irradiated by Ionizing radiation once again. If the sample is irradiated in ice or at liquid nitrogen or liquid helium temperature, the emission of glow may start at any temperature above the temperature of irradiation. Bhasin [1] has made an instrument to measure a glow curve from LNT to about 350 °C. Using this instrument one may see in LiF:Mg, Ti, a strong glow appearing at about −128 °C (Fig. 1.1). The emission of glow may appear in one or more flashes when the sample temperature is raised slowly. The record of light intensity (measured by an instrument such as photomultiplier tube) against time or against temperature is called a glow curve. Usually, however, the intensity is recorded against the temperature. The individual flashes of light in the glow curve record are called glow peaks. In some samples, for example in blue–green variety of fluorite mineral the flashes (glow peaks) can be visually seen coming one after another if the sample is heated in a dark room after a high dose of X or gamma radiations. In some samples even the color of light differs from one flash to other. The color or the spectral property of the glow peak tells about the TL light emitter species, which may be present as an impurity in the bulk sample.

1.2 Early History

The phenomenon of TL was discovered by Sir Robert Boyle [2] who reported to the Royal Society of London on October 28, 1663, the observation of a "glimmering light" when he held a diamond for "a good while upon the warm part of his naked body in the darkness of his bed room". He repeated the observation by using heat from other sources such as a candle, a heated metal and friction. The phenomenon did not find a proper interpretation for about three quarters of a century. The first clear understanding of this observation was provided by Du Fay [3], who showed that this kind of emission was in fact a delayed phosphorescence. Henri Becquerel [4, 5] detected the occurrence of this phenomenon in certain phosphorescent samples toward the end of nineteenth century. He recognized the ability of certain minerals to retain the effect of irradiation from radioactive substances for a long period. Wiedemann and Schmidt [6, 7] used this phenomenon for the first time to detect the cathode rays and UV light. They irradiated a large number of minerals and inorganic compounds with cathode rays and saw intense emission with long storage life in Manganese-doped CaF_2 and in mineral fluorite. These two substances are used for the dosimetry of ionizing radiations even today. The work of Marie Curie [8] on induction of TL in the fluorite mineral by exposure to radium roused great interest in this field. The discovery of X-rays gave further stimulus to the work on TL. During 1925–1928, Wick and co-workers

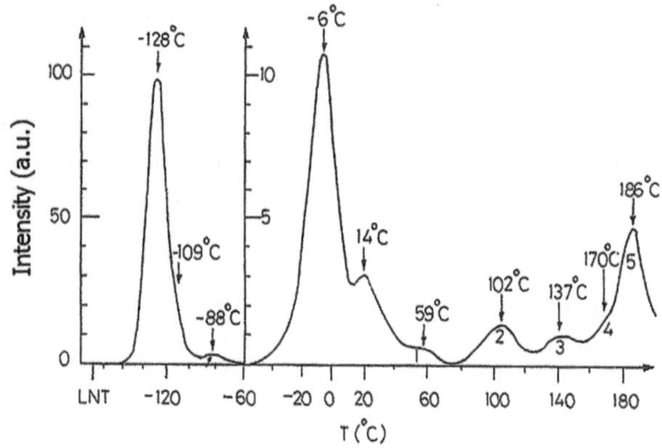

Fig. 1.1 TL glow curve of LiF:Mg, Ti (TLD-100) irradiated at liquid nitrogen temperature. Reprinted from [1]. Reproduced with permission from Elsevier Science

[9, 10] tested the effect of X-rays on the TL of many substances. The effects observed by them included the induction of TL in a number of substances and modification or intensification of it in many others. TL measurements in the form of glow curve were perhaps carried out for the first time in Przibram's [11] laboratory during 1925–1930. His co-worker Urbach [12] reported the TL of alkali halides and attempted to explain them theoretically. The theoretical basis of the TL glow peak was given for the first time by Randall and Wilkins [13]. A modification of this theory was given by Garlick and Gibson [14]. A generalized approach to the kinetics of the process has later been developed by Kelly and Braunlich [15] and Braunlich and Kelly [16].

The present day progress on research in this field owes to the wide application of this phenomenon in the two areas, viz., dosimetry for radiation protection and dating of ancient potteries, ceramics, bricks, and geological sediments. TL has grown basically as an applied branch of science. It was used routinely to measure UV component of sunlight from the early stage of rocket flights [17–19]. The use of different filters including some opaque ones with the phosphor packets on board the rocket produced the direct evidence for the existence of high energy UV components and X-rays in the solar spectrum at high altitudes.

The post-World War II period saw a sudden increase in the sources of ionizing radiations due to the various applications of nuclear energy in military and civil applications. This brought forth the necessity of a simple, reliable, and cheap device for their measurement. Film dosimeters had been in vogue for dosimetry, but were unsatisfactory on many accounts. In late 1940s, therefore began a search for a better and rugged alternative for use in radiation monitoring in military as well as civilian activities. A radiophotoluminescent (RPL) glass dosimeter was first to be developed [20]. It is known that more than a million units of this device were made and used in US Navy [21]. The development of TL phosphors for dosimetry soon followed. In early 1950s, the TL phosphors used in present day dosimetry were discovered. These included LiF by Daniels et al. [22]. The LiF-based TL dose meters were later studied extensively by a group led by Cameron

[23] at the University of Wisconsin. Another phosphor developed early had been CaF_2:Mn which was developed by Ginther and Kirk [24]. Varieties of TL phosphors are now being produced in large scale commercially and are being used throughout the world in radiation protection and medical radiation dosimetry. The rapid progress in the field of TL dosimetry may be gauged from the series of international conferences, which started in 1965 [25] and are being held regularly every 3 years since then. The proceedings of these conferences, which are titled now "Solid State Dosimetry" reflect enormous growth and technical advances. These are an important source of up-to-date information on the science and technology of TL applications.

The potential applications of TL in different branches of human activity were envisaged at an early stage by Daniels. Varieties of applications are summarized by Daniels et al. [26]. Apart from radiation dosimetry, in which luminescence technique has become indispensable, it has been established as an important technique in archaeological and geological dating. Other applications include those in earth and planetary sciences and study of defects in solid crystalline materials. Readers may refer to Aitken [27, 28] for archaeological and geological applications.

1.3 Luminescence

TL is a branch of much wider subject called luminescence. A cursory overview of luminescence phenomena may therefore be useful before taking up the study of the various aspects of TL in detail. Unlike incandescence, luminescence is a cooler and more efficient mechanism of light emission. In luminescence, the light output per unit energy input is much greater than in the case of incandescence. Luminescence is produced by the selective excitation of the atomic or molecular energy levels. Excitation takes the atom or the molecule to a higher energy state. Its fall to ground state results in the emission of light, which is called luminescence. Excitation takes place usually by UV light, but as we would see later in this section, there are other sources which produce excitation. Notwithstanding this simple description it is only a few substances which produce luminescence, even though the exciting light is falling on them throughout the day when the sun is shining. This is because that most of the matter is not made of isolated atoms, with the result that the light energy absorbed is dissipated in increasing the rate and the amplitude of the atomic vibrations which we feel as heat. This competitive process of energy dissipation becomes overwhelming as soon as we depart from the low pressure mono-atomic gases and start dealing with condensed matter. Increasing the vibrational energy of the system raises its temperature and reduces the available energy for light emission. A striking example is the element mercury which is luminescent in the gaseous form but is nonluminescent in the liquid state. This however does not mean that luminescence is impossible in the liquids and the solids. In fact, 75 % of over 3,000 minerals investigated show some kind of luminescence. Some produce luminescence under ultra-violet light exposure and some need X-ray bombardment. The specific characteristic which provides an inorganic solid the luminescent property is the presence of atomic centers whose energy levels are protected from the thermal vibrations of the atoms of the bulk medium. For example the lanthanide rare-earth ions dispersed in calcium salts and in many other crystals and glasses make very efficient luminescent centers. The transitions of inner 4f electrons in them are well protected by the outer shell electrons. In zinc sulfide the copper ions which substitute for a fraction of zinc ions, produce the well known green luminescence. This emission is due to a complex interaction between the impurity and the lattice ions. This involves the coupling of the lattice vibrations with the electronic transition with the result that the emission is a broad structure-less band even at low temperatures. Sometimes there are molecular complexes which emit luminescence. A particularly striking example of protective shielding of molecular energy levels is of the uranyl ion in $UO_2(NO_3)_2\ nH_2O$. The UO_2^{2+} group responsible for luminescence behaves as a separate molecule surrounded by a shield made up of the water molecules of hydration. The three major modes of vibration of $0 = U = 0$ produce three sharp bands in the yellow–green region under UV excitation. At times the luminescent centers consist of ionic vacancies. A very important example of this class is the negative ion vacancy in the alkali halides, in which an electron can be trapped. This is called F center. The electron of the F center has different energy levels like the orbits of a hydrogen atom. Being much less tightly bound than the electrons of the lattice ions, its energy levels fall within the large band gap of the host lattice. This gives rise to absorption bands in the visible range and the corresponding luminescence. While the luminescence of inorganic solids is mostly due to impurity atoms or other lattice defects, the luminescence in organic solids are molecular complexes.

There are a variety of luminescence phenomena observed in nature or in man-made articles. The nomenclature given to these is invariably related to the exciting agent which produces the luminescence. Following description summarizes the main types of the luminescence emission phenomena.

1.3.1 Photoluminescence

This is light emission produced by excitation with the light photons itself. The fluorescent lamp used in household and general lighting is the principal example of this. The 2,537 Å UV radiation from the mercury vapor discharge is absorbed by one of the activator impurities (usually antimony) in the phosphor coated on the inner side of the glass tube. Some of this energy is transferred by resonance to a second impurity (usually manganese). By adjusting the relative concentrations of these activator impurities one can produce desired modification in the color of the light. There are a large variety of organic and inorganic phosphors which are used in consumer items like in road and traffic signals, displays, laundry whiteners, etc., in addition to the host of those used in industrial and scientific applications. One of the high technology subjects called light amplification by stimulated emission of radiation (LASER) is a kind of photoluminescence in which emission is coherent.

1.3.2 Fluorescence and Phosphorescence

The phenomenon of Photoluminescence may be divided into two types namely fluorescence and phosphorescence. These light emission processes may be understood in a simple way by using the energy level diagrams. After an atom or molecule is excited to an upper state by absorption of a light photon, it can return to the ground state in a number of different ways. The simplest way is that the atom returns to its ground state by the re-emission of a light photon of same energy as was absorbed. This is called resonance fluorescence. However, there may be other energy levels between the ground state N and the excited state E. In Fig. 1.2 are shown also the energy levels close but immediately above N and above E, such that the system can be easily raised to these levels by the transfer of thermal energy from the surrounding atoms. The different transitions involving these levels are shown in the Fig. 1.2. The transitions from E to any of the levels A, B, … give rise to fluorescence light in which the energy of the photons is smaller than the exciting photon energy. This energy difference is called the Stokes shift. Fluorescence emissions of photon energy lesser than the excitation energy are most common. There are however exceptions. The presence of energy levels N′ and E′ immediately above N and E, respectively often gives rise to fluorescence light of energy greater than the exciting light. Such fluorescence emission is called anti-Stokes fluorescence. When the ground state N is raised to N′ by the thermal energy before the exciting light photon is absorbed, the emitted fluorescence has an energy (transition E → N) which is higher than that of the exciting light (transition N′ → E). Similarly when the thermal agitations raise the system to E′ after it has absorbed the excitation energy, the resulting emission (transition E′ → N) again has an energy greater than that of the exciting light (transition N → E). For various transition schemes in fluorescence and phosphorescence, the reader may refer to Pringsheim [29] or any other book on Luminescence.

Phosphorescence may be defined as the emission of luminescence after an electron is brought to a metastable state subsequent to the excitation of the system. A metastable state M is an energy level from which transitions to any lower energy levels are forbidden. If a system brought to its metastable state is completely unperturbed (for instance, an atom in a highly rarefied atmosphere) it would remain in such a state for a relatively long period. Absorption transition from the ground state to M also is forbidden. It can however, be reached indirectly. For example after the system has attained the state E consequent to excitation, it can

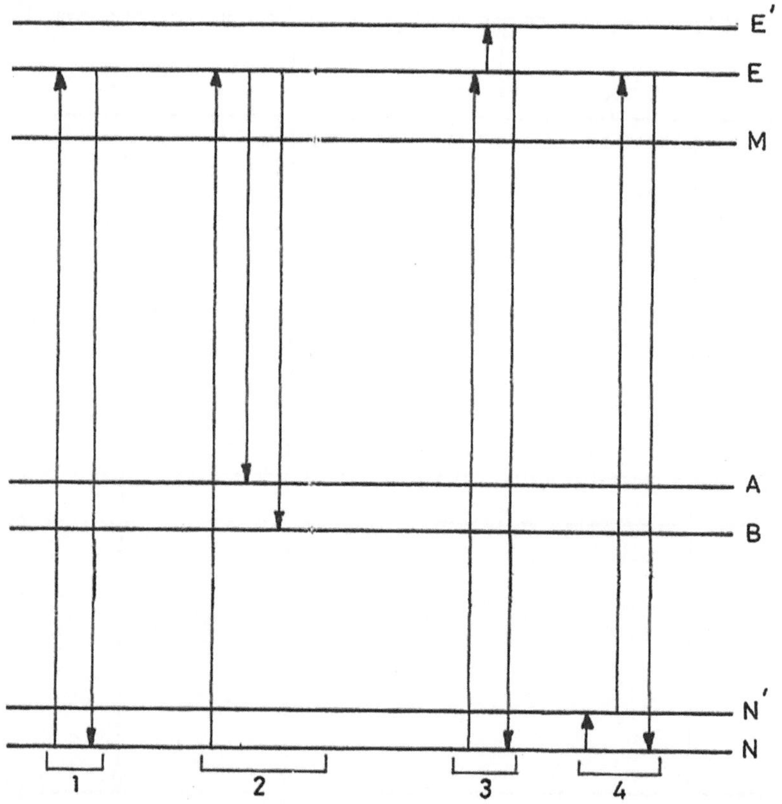

Fig. 1.2 Energy level diagram illustrating the fluorescence emissions. All upward transitions show absorptions (excitations) and all downwards show the emissions. *1* resonance fluorescence, *2* fluorescence with Stokes shift, *3* and *4* fluorescence with anti-Stokes shift

fall to M by the transition E → M (Fig. 1.3). If M is separated from E by a small energy gap and if the system is in thermal equilibrium with its surrounding medium, it can pick up thermal energy to return to state E. From there transitions can take place as described earlier in the case of fluorescence. The phosphorescence spectrum thus would contain the emissions which are present also in fluorescence. The thermal excitation from M to E hastens the depopulation of the state M. The emission intensity as well its decay time thus changes with even small change in temperature. Phosphorescence thus is sensitive to changes in temperature in contrast to fluorescence. Conventionally emissions with decay time shorter than 10^{-8} s are assumed as fluorescence and those with decay time greater than that are taken as phosphorescence. However, it is the temperature dependence of decay time which is the real test to differentiate phosphorescence from fluorescence.

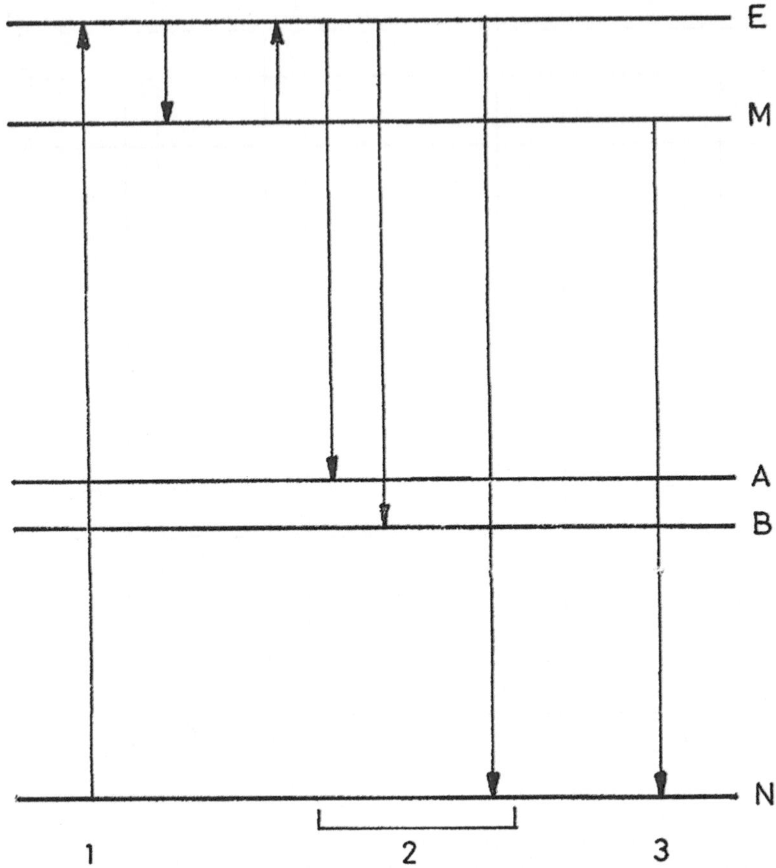

Fig. 1.3 Energy-level diagram illustrating the phosphorescence emissions. *M* is a metastable state, *1* excitation transition, *2* temperature sensitive phosphorescence. Down and up transitions between E and M are thermal energy (*vibrational energy*) exchanges, and *3* long-lived phosphorescence

1.4 Luminescence in Condensed Matter

The type of phosphorescence described above usually applies to the atomic and molecular systems. In solids and liquids there are other mechanisms which lead to luminescence with a long decay time. In these cases, the excitation energy travels from the site of absorption to the site of emission, the two being separated in space in the medium. This transfer of excitation energy may be divided into two types: (1) excitonic transfer and (2) transfer by electrons and holes. A hole means nothing else but a vacancy of an electron. An exciton is a bound electron-hole pair which can move like a free particle. It may also be visualized as a mobile excited state in the lattice. Typical absorption lines due to the exciton transitions are seen in the low temperature UV absorption spectra of alkali halides. Examples of these are given by Knox and Teegarden [30]. When a mobile exciton comes in contact with a defect in the lattice, it may transfer its energy to the latter, which in its turn goes to its excited state. If the defect relaxes by a radiative transition, we see the emission of light. This type of process is well known in the impurity doped alkali halides such as NaI (Tl). The exciton in this case is the mobile excited state of the iodide ion and the impurity is Tl^+ ion present substitutionally in place of Na^+.

The other type of energy transport from the site of absorption to the site of emission involves ionization. The term "ionization" means ejecting out an electron from its normal position. The vacancy left behind is called a hole. For example, an ionization event in NaCl crystal means ejecting out an electron from Cl^- of the NaCl lattice. The Cl° left behind may be described as a hole in the valence band. Unlike exciton transport in which the electron and hole are bound together, in this case the two entities move independent of each other after separation from the site of ionization. When we talk of a moving hole like a moving electron, we mean the electron vacancy is hopping from one atom or ion to another in the lattice. In material sense, it is the hopping of the electron in the opposite direction. Any radiation with photon energy greater than the band gap of the solid can cause ionization. Some of the electrons and the holes liberated during ionization get caught up in some defect locations in the lattice of the sample material. For example, an electron may be captured by a divalent cationic impurity present substitutionally in a monovalent lattice point, or it may be captured at a negative ion vacancy site. Similarly, a hole may be captured at a location where positive charge is lacking or a negative charge is in excess, for example at a positive ion vacancy or at an interstitial negative ion. These locations where electrons and holes are caught up are called the "traps". The binding energy of the trapped charge carrier is called the "trap depth" or the "activation energy" of the trap. Traps of stability in the range of seconds or minutes at an ambient temperature lead to phosphorescence afterglow. Some of the traps are so stable that at ambient environmental temperatures the trapped charges may stay in them for hundreds or thousands of years. These trapped charges may be made free by providing stimulating energy (activation energy) by means of heat or light photons of energy equal to or greater than the binding energy of the trapped charge. The recombination of these free charges at a suitable site gives rise to luminescence emission. When the stimulating agent for liberating the trapped charges is heat, the emitted luminescence is called *TL*. When it is the light photon, the emission is called *optically stimulated luminescence* (OSL). Both excitonic as well as the electronic processes of energy transport are important. While the latter is the main process during TL induction as well as during thermal stimulation, the excitonic process is often involved in recombination and emission processes in the case of certain glow peaks of alkali halides.

1.4.1 *Optically Stimulated Luminescence*

The two phenomena namely the OSL and the TSL are like the twin sisters descending from the same parentage. The induction or pre-excitation process described briefly in the preceding paragraph is common in both. It is the stimulating agent which is different in the two. When the stimulating agent for liberating the trapped charges is heat, the emitted luminescence is called TSL or TL. When it is the light photons the emission is called the OSL. The phenomenon TSL is a kind of phosphorescence. In the last decade, the OSL phenomenon has become more popular than TSL in its application in radiation dosimetry and dating of archaeological and geological samples. The TSL technique however, continues to hold its place in the applications as well as in the study of materials used in these applications. Yukihara and McKeever [31] have given a detailed account of the theory and the applications of OSL in a recently published book.

1.5 Cathodoluminescence

When excitation is done by electron beams generated at the electrical cathodes, the emission produced is called *cathodoluminescence.* The screens of cathode ray tubes and earlier version of televisions glow by this kind of emission. In cathode ray tubes, zinc and cadmium sulfide phosphors are used. Production of phosphors for TV screens is a very specialized technique, which requires variety of colors and their appropriate persistence to smoothen out the flicker of the scan.

1.6 Radioluminescence

When the excitation energy is provided by X-rays or nuclear radiations such as α, β, and gamma rays, the resulting luminescence is called *radioluminescence*. Luminescence produced by X-rays has also been named as X-ray excited optical luminescence (XEOL) to differentiate it from X-rays fluorescence (XRF). In both of these, the excitation is by X-rays but the emission is in the form of light photons in the former and it is in the form of X-ray photons in the latter. Both these techniques are very useful in trace analysis of elemental impurities. The mechanism of light emission in radioluminescence and cathodoluminscence is same because in both the cases it is the electrons, incident primarily from cathode rays or the secondary electrons produced after the first interaction of the incident X-ray photon or nuclear particle, which cause the excitation of the luminescent species in the bulk sample. There are varieties of inorganic minerals which show strong luminescence under excitation by X-rays and nuclear particles, although they show no luminescence under UV excitation. The luminescent dials of watches of yester years and of many instruments shine in darkness due to the luminescence excited by beta particles of tritium added to the ZnS phosphor. These are no more in vogue now because of use of radioactivity in them.

1.7 Scintillation

This phenomenon is similar to radioluminescence. It is given the name scintillation, because it is used as a technique to detect individual light pulses generated by the incidence of each X or gamma ray photon or a nuclear particle. Such light pulses are called scintillations, since like a spark they are very short-lived. Thallium-activated sodium iodide is a well-known scintillation detector used for gamma ray spectrometry. The intensity of the scintillation (Light pulse) is directly proportional to the incident gamma ray photon energy when it is totally absorbed. The measurement of the pulse intensity, therefore, provides the means for knowing the gamma ray energy. Anthracene and special type of plastics are used for beta ray measurements. ZnS phosphor is used in alpha ray counting. There are liquid scintillators which are used in low energy beta counting like those from carbon-14 and tritium.

1.8 Electroluminesce

Application of electric fields can produce luminescence in many phosphors. The mechanism of electroluminescence is complicated. Unlike in cathodoluminescence or roadioluminescence, electric fields by themselves cannot kick the electrons free from their valence states. It is believed that steep voltage gradients are generated across the crystalline boundaries or other crystal defects, when electric fields are applied. Electrons reaching these regions get sufficiently accelerated to attain enough energy to ionize the atoms. In common practice, alternating voltages are applied across thin layers of the phosphors. There is another type of electroluminescence known as injection luminescence. In this electrons are injected from an external supply across a semiconductor p-n junction. On applying a DC voltage across the junction, such that the electrons flow to the p region, luminescence is produced by the electron-hole recombination in that region. The light emitting diodes (LED) which are now commonly used as display devices in many scientific instruments are based on this principle. LEDs are now being developed as general purpose light sources so that these may replace the mercury vapor based fluorescent lamps. Bhushan [32] has given a comprehensive review on electroluminescence.

1.9 Chemiluminescence

Certain chemical reactions produce luminescence. The oxidation of white phosphorous in air is the best-known example of this. Some molecules which are formed in high energy electronic states radiate out the energy as light, while the bulk material remains at ambient temperature. Chemiluminescence is, however not a common accompaniment in all chemical reactions, because the amount of energy released even in the exothermic reactions is not sufficient to cause electronic excitation, which needs a couple of eV energy. Secondly not all chemical molecules are capable to produce luminescence. *Lyoluminescence*, which is caused during the dissolution of certain compounds which have been bombarded by X-rays beforehand is a kind of chemiluminescence. A well-known example is the case of X-irradiated NaCl which emits a flash of light when quickly dissolved in water. The electrons trapped in the NaCl lattice due to the X-ray bombardment, when freed during dissolution are believed to excite the luminescence.

1.10 Bioluminescence

Biochemical reactions inside the cells of the living organisms can produce electronic excited states of the biomolecules. Fire flies, glow-worms, some bacteria, and fungi and many sea creatures (Such as planktons) both near surface and at great depths are the striking examples of luminescence in living beings. The chemical reactions are the enzymic oxidations. For example, it has been explained as the oxidation of luciferin in the presence of enzyme luciferase being responsible for the glow of the firefly. Similarly, sea crustacean and the luminous bacteria also give a glow, even though the chemical nature of the luciferin in each of these is said to be quite different from each other.

1.11 Triboluminescence

A large number of inorganic and organic materials when subjected to mechanical stress, emit light which is called triboluminescence. It has also been named as mechano-luminescence by some authors. It has been observed that all piezoelectric crystals exhibit triboluminescence. The spectra of triboluminescence light are similar to those of photoluminescence in many substances. The excitation is due to electrification of the cleavages and cracks. The discharge in the surrounding gas also adds to the light produced in the stressed crystal. Chandra [33] has summarized the mechanism of this emission in different substances.

1.12 Nature's Luminescence Wonders

There are many fascinating Phenomena in nature which are related to luminescence emission. The glow worms, fire flies, some bacteria and fungi, and varieties of sea creatures as self luminescent creatures have been mentioned above. The polar lights called auroras are another spectacular phenomena related to luminescence. It is believed that these colorful flashes of light are produced by the excitation of the atmospheric oxygen and nitrogen by high energy cosmic ray particles particularly during the solar flares.

References

1. B.D. Bhasin, Integrated system for the study of TL, TSC and related phenomena from liquid nitrogen temperature to 800 K. Nucl. Tracks **10**(1–2), 253 (1985)
2. R. Boyle, in *Register of the Royal Society*, vol. 1663 (London, 1663), p. 213
3. C.F. Du Fay, Hist. de Acad. Roy de Sci. de Paris **1738**, 347 (1938)
4. H. Bacquerel, C.R. Acad. Sci. (Paris) **129**, 912 (1889)
5. H. Bacquerel, C.R. Acad. Sci. (Paris) **144**, 619 (1907)
6. E. Wiedemann, G.C. Schmidt, Ann. Phys. Chem. **54**, 604 (1895)
7. E. Wiedemann, G.C. Schmidt, Ann. Phys. Chem. **56**, 237 (1895)
8. M. Curie, Radioactive substances (English translation), Doctoral thesis, Faculty of science, Paris, referred by S.W.S. Mckeever in Thermoluminescence of solids, Cambridge University Press, Cambridge, 1984
9. F.G. Wick, M.K. Slattery, J. Opt. Soc. Amer **16**, 398 (1928)
10. F.G. Wick, S.B. Acad. Wiss Wien IIa **139**, 497 (1938)
11. K. Przibram, in *Irradiation Colours and Luminescence*, (Pergamon Press London, London, 1956)
12. F. Urbach, Wein. Ber. IIa, 363 (1930)
13. J.T. Randall, M.H.F. Wilkins, Proc. Roy. Soc. (London) ser. A, **184**, 365 (1945)
14. G.F.J. Garlick, A.F. Gibson, Proc. Phys. Soc (London) **60**, 574 (1948)
15. P. Kelly, P. Braunlich, Phys. Rev. B **1**, 1587 (1970)
16. P. Braunlich, P. Kelly, Phys. Rev. B **1**, 1596 (1970)
17. K. Watanabe, M.K. Slattery, J. Opt. Soc. Amer. **16**, 398 (1928)
18. T. Lyman, Phys. Rev. **48**, 149 (1935)
19. R. Tousey, K. Watanabe, D. Purcell, Phys. Rev. **83**, 792 (1951)
20. K. Becker, *Solid State Dosimetry* (CRC Press Cleveland, Ohio, 1973)
21. J.H. Schulman, W. Shurcliff, R.J. Ginther, F.H. Attix, Nucleonics **11**(10), 52 (1953)
22. F. Daniels, in *Luminescence Dosimetry*, ed. by F.H. Attix. Proc. First Int. Conf. Luminescence Dosimetry, US AEC, Div. Techn. Information Report Conf. 650637 (1965), p. 34
23. J.R. Cameron, N. Suntharalingam, G.N. Kenny, *Thermoluminescence Dosimetry* (University of Wisconsin Press, Madison, 1968)
24. R.J. Ginther, R.D. Kirk, J. Electrochem. Soc. **104**, 365 (1957)
25. F.H. Attix (ed.), in *Proceedings of the First International Conference luminescence Dosimetry,* USAEC Div. Tech. Information, Report Conf. 650637 (1965)
26. F. Daniels, C.A. Boyd, D.F. Saunders, Science **117**, 343 (1953)
27. M.J. Aitken, *Thermoluminescence Dating* (Academic Press, London, 1984)
28. M.J. Aitken, Archaeological involvements of physics. Phys. Lett. **40C**, 279 (1978)

29. P. Pringsheim, *Fluorescence and Phosphorescence* (Interscience Publications, New York, 1949), p. 3
30. K.S. Knox, K.J. Teegarden, in *Electronic Excitations of Perfect Alkali Halide Crystals, in Physics of Colour Centers*, ed. by W.B. Fowler, (Academic Press, London, 1968), p. 1
31. G. Eduardo Yukihara, W.S. McKeever Stephen, *Optically Stimulated Luminescence Fundamentals and Applications* (Wiley, New York, 2011)
32. S. Bhushan, Nucl. Tracks **10**, 215 (1985)
33. B.P. Chandra, Nucl. Tracks **10**, 225 (1985)

Chapter 2
Induction of Thermoluminescence

Abstract The essential condition for induction of thermoluminescence (TL) in a material is the presence of suitable defect centers. This chapter summarizes well-known type of point defects in the inorganic solids. These serve as simple examples of defect centers, the types of which may act as traps and recombination centers. However, the actual structure of the defect centers in different TL phosphors is not simple to find out, neither is it necessary for the application of TL in areas like dosimetry and archaeological and geological dating. Defect centers may be created by doping the material with suitable impurities and thermal treatments like heating at a high temperature and quenching. X-rays and nuclear radiations being the principal sources which induce the TL, their interaction with matter is summarized in this chapter. The interactions are discussed in the perspective of the TL induction by different types of radiations.

2.1 Essential Conditions for TL Induction

As discussed in Chap. 1, thermoluminescence (TL) in a sample may be described as a two-step process: (1) induction (excitation) of TL by exposure to ionizing radiations, (2) stimulation of TL emission by heating subsequent to exposure. The necessary condition for the induction of TL in a material is that the concerned material should contain certain types of defects in its regular structure. This means that very pure and defect free materials would not show TL. These defects should be capable of capturing electrons or holes during exposure to ionizing radiations. Further, the captured charges should be retained in that state until the sample is heated to an appropriate temperature to read the TL. These defect structures may be called the TL centers. The term "center" is used to designate these structures because these are present as isolated entities and are not a part of the regular lattice, which is a continuous repetition of the unit cell of a crystalline material. The function of irradiating the material with ionizing radiations is to produce free electrons and holes in abundance by the interaction of the incident radiation with

C. M. Sunta, *Unraveling Thermoluminescence*,
Springer Series in Materials Science 202,
DOI: 10.1007/978-81-322-1940-8_2, © Springer India 2015

atoms of the bulk sample. Majority of the electrons and holes undergo recombination almost instantaneously during irradiation and in the process one may see the radioluminescence. Some of these free electrons and holes diffuse in the lattice and are ultimately captured (trapped) in the defect centers. Irradiation may also produce the defects in addition to producing ionization. This will be discussed in the following sections. In order that the trapped charge carriers (electrons and holes) are able to produce luminescence emission during thermal stimulation, it is necessary that the sample material contains also such defect species which capture the charge carriers liberated from the traps during heating and in the process undergo electronic transitions leading to the emission of light. These species are called *recombination centers* or luminescent centers. In the following, we briefly describe some of the well-known defect types primarily in alkali halides. These are given as mere examples, since the defect types in various TL materials depend on the material itself and one may or may not be able to know their exact structure. Notwithstanding this difficulty these materials are useful in applied disciplines.

2.2 Defect Centers in TL

The defect centers may be divided into two categories: (1) those inherently present in the material and (2) those produced by external means, such as by purposely doping the sample with impurities. The inherently present defect centers may appear simply due to thermodynamic reasons. They may appear also due to the presence of trace impurities, which may be difficult to be removed during the purification process. The equilibrium concentration of defects at any temperature due to thermodynamic reason may be calculated from statistical mechanics. The concentration of these defects at an ambient temperature may be increased by taking the sample to a higher temperature and then cooling it suddenly (quenching). In this way, the increased number of defects present at higher temperatures may be frozen. This method is quite common in the preparation of TL samples. When an ion or atom from its regular lattice position slips into an adjacent interstitial position, the defect formed is called Frenkel defect. It is thus a vacancy—interstitial pair. Cations due to comparatively smaller ionic radii than those of anions are more likely to form Frenkel defects in alkali halides. Cation and anion vacancy pairs are called Schottky defects. It is observed that Schottky defects are most numerous in ionic crystals. The concentrations of Frankel (n_i) and Schottky (n_v) defects in thermodynamic equilibrium at temperature T K are given by Schulman and Compton [1] as

$$n_i = (ANN_i)^{1/2}\exp(-W_i/2kT) \tag{2.1}$$

$$n_v = BN\exp(-W_s/2KT) \tag{2.2}$$

where

n_i number of ions in interstitial positions
n_v number of vacant lattice sites
N total number of lattice points in the ideal crystal
N_i total number of interstitial positions
W_i energy required to form a Frankel defect
W_s energy required to form a Schottky defect. A and B are constants

In alkali halides estimates of W_i and W_s are; $W_i = 2.9$ eV for Na^+ in NaCl and $W_s = 1.9$–2 eV for NaCl, KCl, and KBr. This shows that there will be more Schottky defects in alkali halides, since the energy requirement is less for the formation of these than for the Frankel defects. The *F* center, most commonly observed point defect is a negative ion vacancy, which has captured an electron. Lattice vacancies may also be caused by higher valency impurity atoms present substitutionally in the lattice. A divalent cation impurity in a monovalent lattice for example would induce the formation of a cation vacancy in the lattice to maintain charge neutrality. Similarly anionic vacancies may arise if for example an O^{2-} ion substitutes for a halide ion in alkali halides. The impurities (*I*) and the vacancies (*V*) tend to associate to form *I*–*V* pairs at low temperatures. These pairs may further associate to form dimers, trimers or even larger clusters. In the following are listed examples of the defects under two categories namely those which trap electrons and those which trap holes.

2.2.1 Defects Which Trap Electrons

Often these consist of negative ion vacancies. A negative ion vacancy has a net positive charge. In alkali halides, for example, the negative ion vacancy, being surrounded by six cations, the vacant space is in the positively charged field. It can thus easily capture an electron. The electron is not stationary in the vacancy. ESR studies have shown that the trapped electron moves in an *S* type orbit. In fact the *F* center has been likened to a hydrogen atom, in which the electron is bound to a strong Coulombic central force. Like hydrogen atom it has finite number of bound states. The orbitals may overlap with those of other nearby *F* centers, if present. This leads to the formation of aggregates of *F* centers. These are named as *M*, *R*, *N*, etc., centers [2]. The various electron trap centers arising from the negative ion vacancies are listed in Table 2.1 and are shown in Fig. 2.1. In addition to those listed therein, varieties of electron traps are produced also by cationic substitutional impurities. For example, divalent impurities like Pb^{2+}, Mn^{2+}, Mg^{2+}, etc., in alkali halides act as strong electron traps. Similarly trivalent rare earth impurities (RE^{3+}) in alkaline earth compounds may form the main electron traps [3–5]. These may compete with *F* center formation at negative ion vacancies and play prominent role in the emission of TL. The extra positive charge of the

Table 2.1 Defect centers in alkali halides [2]

Name of the center	Description
(a) *Electron trap centers*	
F	An electron at a negative ion vacancy. Its name originates from the word farbzentren (farbe means color and zentren means center in German language)
F'	Two electrons associated with a negative ion vacancy
F_A	It is an F center adjacent to a substitutional impurity ion of same group, but with lesser atomic number, for example Li^+ or Na^+ in place of K^+ in KCl and KI
F^+	A negative ion vacancy
M	Two neighboring F centers in [110] axis
R	Three neighboring F centers forming an equilateral triangle in a [111] plane
Z	These centers are in a way similar to F_A centers except that the nearby impurity is of higher valency. There is a series of Z centers named Z_1, Z_2, Z_3, etc. Z_1 is an F center with a cation impurity of higher valency and a cation vacancy nearby. Z_2 center is an F' center together with a substitutional cation impurity of higher valency. Z_3 is an ionized Z_2 center. Z_0 is a negative ion vacancy near a substitutional divalent impurity
(b) *Hole trap centers*	
V_K	One hole (absence of one electron) shared by two adjacent anions. It is a molecular ion, for example F_2^-, Cl_2^-, etc., in alkali halides. These centers are usually observed only at low temperatures. In alkali halides, these are stable only below about 80 K. The subscript K in V_K stands for Kanzig who discovered this center
V_{KA}	It is analogous to F_A. If one of the nearest neighbor lattice site of V_K center is occupied by an alkali impurity of lesser atomic number, for example Li^+ in place of Na^+ in NaF, it is called V_{KA} center
H	It is similar to V_K except that the hole has interaction with two more nearest halide ions. The hole is thus shared between four halogen ions in a straight line. It is represented by X_4^{3-}. This center is stable up to 130 K in LiF
H_A	It is analogous to F_A and V_{KA}. It is an H center adjacent to which there is a substitutional cation impurity of the same group as the host cation but of lesser atomic number
V_F	It is an antimorph of F center, it is a V_K center adjacent to a cation vacancy. Its absorption overlaps the V_K band but it is more stable than V_K

substitutional impurity in the cationic lattice is neutralized by electron capture in these during irradiation. This reduces the valence state of the impurity to render it similar to that of the host lattice ions.

2.2.2 Defects Which Trap Holes

Like the anionic vacancies which form the electron traps, the cationic vacancies are potential sites for trapping the holes. A hole located at the site of a cationic

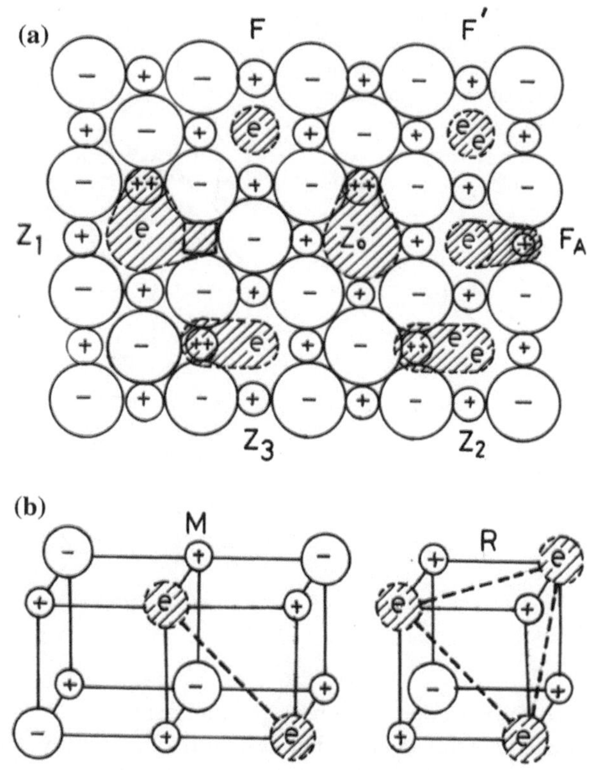

Fig. 2.1 **a** Electron trap type defect centers. The vacancies in which electrons are trapped are shown by hatched parts. The *F*, *F*′ F_A and four types of *Z* centers. **b** The aggregates of *F* centers named *M* and *R* centers

vacancy may be considered as an antimorph of an *F* center. However, whereas the *F* center has an *S*-type orbital, the hole in such a center has a *P*-type state in alkali halides. In the physical sense, one may understand the hole as the absence of an electron from the anion which is adjacent to the cationic vacancy. In alkali halides, the absence of the electron is in the closed *P* shell of the halide ion. In fact the hole is shared by two neighboring halide ions adjacent to the cation vacancy (V_F center in Fig. 2.2). Substitutional Na^+ and O^{2-} in cationic and anionic sites respectively in CaF_2 lattice are understood to act as hole traps. Similarly interstitial F^- ions also offer as hole traps in CaF_2 [3–5]. Some typical electron and hole type centers in alkali halides are listed in Table 2.1 (see also Fig. 2.2).

Some types of hole centers are shown schematically in Fig. 2.2.

2.3 Defect Generation by Irradiation

The incidence of ionizing radiations may produce defect centers in two ways: (1) ionization of a lattice ions leading to their displacement into interstitial position. (2) elastic collisions of heavy particles like alpha, energetic ions, and fast neutrons.

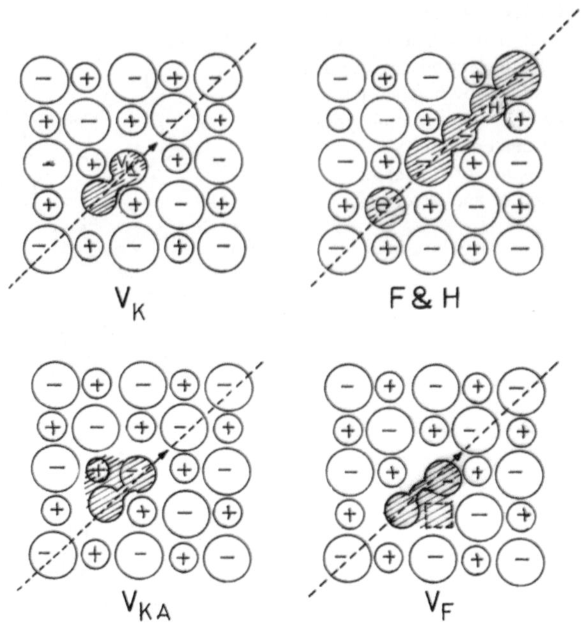

Fig. 2.2 Hole trap type centers. The vacancies where holes are trapped are shown as hatched parts in the same way as in Fig. 2.1

The former appears to be a predominant way of causing vacancy and interstitial related defects in alkali halides, particularly at low temperatures. Defect formation by ionization generally takes place in halide lattice. This is because most often it is the valence electrons, which are ejected out during ionization. It is so because the valence electrons are energetically least tightly bound. When an electron is removed from a halide ion, a halogen atom is formed in the anionic position. The halogen atom, being smaller than the halogen ion in size and not bound electrostatically to the lattice (because it is neutral), a relaxation of lattice takes place around the halogen atom. This sets the nearest neighbor ions into motion. As a result of this motion, the unbound neutral halogen atom is most likely candidate to diffuse away from the lattice site and occupy an interstitial position. This interstitial halogen atom, by sharing the deficiency of one electron with three other halide ions forms the so-called H center. The vacancy left behind in the anion lattice position forms an F center by trapping an electron (Fig. 2.1). A refined mechanism for the formation of an *F* and *H* center pair, which is now most widely accepted is as follows [2]: The halogen atom left behind after ionization of the halide ion, gets associated with its neighbor halide ion forming a halide molecule (X_2^-). This singly charged halide molecule in which one electron is shared between the two halogens is called a V_K (Fig. 2.2) type hole center or a "self trapped hole." It is stable only at low temperatures. The two halide ions in this molecule are displaced from their normal lattice positions because of having smaller spacing between them than the normal negatively charged halide ions. At higher temperatures, a displacement sequence is initiated which ultimately leaves a vacancy in the anionic position together with the formation of an *H* center. The H center is referred also as 'crowdion'. It is a crowd of four halogens distributed over three lattice positions (X_4^{3-}). This is a hole center in which the hole is shared between four halogens in the same way as one hole is shared between two halogens in a V_K center. Recent treatments of this model propose appreciable migration of V_K center (through excited state, i.e., excitonic diffusion) before the final relaxation into *F* and *H* centers takes place [6, 7]. There is a strong probability that *F* and *H* pairs are stabilized near the impurities [7]. In real materials there may be a number of modifications of these basic centers in different substances. For example, the *F*, V_K or *H* centers are all modified generally with enhanced stability when these are located adjacent to substitutional cationic impurities. Monovalent alkali impurities such as Li or Na in KCl, KBr, KI, etc., which have trapped interstitial halide ions give rise to the so-called H_A centers. Similarly *F* centers stabilized near divalent substitutional impurity sites are called *Z* centers [8]. These modified forms may be involved in the TL process of individual materials.

Different types of centers irrespective of whether electron or hole type, may act as traps or recombination centers. Convention has it that those centers from which the charge carrier is liberated during thermal stimulation are called the traps and those with whom the thermally liberated charge combines are called recombination centers. Recombination centers are more stable than their counterpart traps. It is usually presumed that the luminescence emission occurs from the recombination center. This, however, may not be always so. Quite often the energy released during recombination is transferred to a neighboring impurity ion, which ultimately undergoes the light emitting transition. The light emitting entity may sometimes be identified from the wavelength spectrum of the emitted light. Examples of such cases are the rare earth doped calcium based TL phosphors [4, 9, 10]. Identification of either traps or recombination centers has not been successful in many TL phosphors used even in routine dosimetric applications. Correlative studies using optical absorption and EPR investigations [11, 12] together with the recording of TL glow curves and their wave length spectra may be useful for the identification of the traps and the recombination centers of the TL materials.

2.4 Interactions of Radiations

For the purpose of discussing the interactions of the incident radiations, which lead to the induction of TL in insulating solids like phosphors and minerals we may divide the radiations into the following four categories: (1) Heavy charged particles, such as alpha particles and protons. (2) Light charged particles, which essentially consist of electrons. These are also named as cathode rays or beta rays depending on their origin. Positrons also are included in the same category. These have same mass as electrons, but are positively charged. (3) Electromagnetic radiations, i.e., gamma and X-rays. This category includes the vacuum ultraviolet rays. (4) Neutral particle radiations namely neutrons. There are other particle radiations like mesons which are present in cosmic rays but are not important for the present discussions.

2.4.1 Heavy Charged Particles

Under this category, let us consider alpha particles and protons. Alpha particles consist of two protons and two neutrons and are like helium nuclei. These thus have double the charge of protons and are four times as heavy. The protons are 1,840 times more massive than electrons and carry one unit of positive charge. The interaction of these heavy charged particle radiations can be divided into two major types: (1) inelastic collisions with atomic electrons. (2) elastic collisions with the nuclei of the atoms of the target material. The former is the predominant mechanism by which these charged particles lose their kinetic energy in a target sample. Being positively charged alpha particles and protons exert a coulomb force of attraction on the electrons and pull some of these away from their binding with the parent atoms. Being relatively massive, alpha particles have short range and consequently high specific ionization. Some of the collisions of the alpha particles with atomic electrons are so hard that the ejected out electron produces its own track of interactions with electrons of other atoms. These secondary energetic electrons are called delta rays. Alpha particles from usually available radioactive sources (4–7 MeV) produce a few tens of thousand ion pairs each when incident into a solid insulating medium. The excitation of TL is caused by these ions (electrons). Alpha particles and protons produce densely ionized isolated tracks in the sample volume with the result that the ionization is nonuniform with a large part of the sample volume remaining un-irradiated. The efficiency of TL induction by the heavy charged particles is therefore low. For alpha particles, it is only 4 % of that by β, X and gamma rays for a given dose [13].

2.4.2 Light Charged Particles

Under this heading we will consider the beta particles, energetic electrons, and positrons. The effect of both the electrons and positrons on the matter is similar except for the fact that the positron ultimately annihilates itself by combining with an electron giving rise to a pair of 0.51 MeV gamma photons. Let us consider the direct interaction between an incident electron and the target matter. Suppose an electron with velocity v is moving in the matter. The electron has a radial electric field. As it moves through the interstitial spaces of the atoms, it exerts Coulomb force of repulsion on the orbital electrons of the target medium. The energy imparted to the orbital electrons depends on the length of the time the incident particle spends in the vicinity of the orbital electron and the approach distance. If enough energy is transferred, the orbital electron gets ejected out. Some of the interactions may occur at close distances, with the result that the bound orbital electron may be ejected with sufficiently high energy. Such electrons are called delta rays. These dissipate their energy in the same manner as the original incident beta particle. The energetic beta particles lose energy in yet another manner: when a beta particle passes through the Coulomb field of a nucleus, it experiences a strong force and loses its energy by the so called process of bremsstrahlung emission. The bremsstrahlung radiations are X-rays produced by the "braking action" of coulomb field of the nucleus on the passing electron. This process of energy loss becomes important for beta energies above several MeV and is greater for target elements of high atomic number, since higher the atomic number, greater is the nuclear charge and hence greater is the braking action. The bremsstrahlung radiations interact in the sample as discussed later for X and gamma rays. Unlike the alpha particles beta particles do not have a straight line path. Particularly at lower energies the path of an electron is apt to be tortuous due to multiple scattering in the Coulomb field of electrons or nuclei in its path. The ionization produced in the sample is thus uniform for a beta ray beam.

2.4.3 Electromagnetic Radiations (X and Gamma Rays)

X-rays are produced by the stoppage of fast moving cathode ray electrons in a target, which is referred above as bremsstrahlung radiation. Gamma rays are emitted from the nuclei of the radioactive elements during their disintegration. There is no basic difference between the properties of X and gamma rays notwithstanding the difference in their origin. When incident on a material these interact in the following three ways: (i) Photoelectric effect (ii) Compton effect and (iii) Pair production. In photoelectric effect, an electron is knocked out usually from the inner orbit of the target atom. The knocked out electron which is called photoelectron carries with it almost all the energy of the incident photon. The kinetic energy T of the photoelectron is given by

$$T = E - W \tag{2.3}$$

where E is the incident photon energy and W is the binding energy of the atomic orbit from which the electron is removed. W is also called the work function. The cross-section σ of photoelectric interaction is approximately given by $\sigma = AZ^4/E^{3.5}$ where A is a constant and Z is the atomic number of the target material. The strong dependence of photoelectric effect on Z and E is evident from this equation. In LiF phosphor, for example, most of the interactions would take place in Fluoride ion (Z of F being much higher than that of Li).

Compton Effect is a scattering phenomenon. In this, the incident X or gamma photon knocks out an electron from an orbit of the target atom by imparting part of its energy to the orbital electron. The incident photon is scattered, carrying with it the remaining energy. The scattered photon may undergo another Compton collision or it may interact photoelectrically, in either case imparting energy to yet another electron. The Compton scattering cross-section is a complicated function of the energy of the incident photon and the angle of scattering. It is independent of the atomic number Z of the target material. However, the Compton mass absorption coefficient (given in units of $cm^2\ g^{-1}$) depends on the density of the electrons. It is thus greater for lighter elements since the number of electrons per atomic number is higher for lighter elements.

In the third type of interaction, namely the pair production, the X or gamma photon gets converted into a pair of positive and negative electrons when the photon passes near the nucleus of an atom. This effect takes place only when the incident photon energy is more than 1.02 MeV, which is the sum of the mass equivalent of one electron plus one positron. The excess energy of the incident photon over 1.02 MeV is carried by the electron and the positron in the form of their kinetic energy. The positron after spending its kinetic energy recombines with an electron producing two gamma photons of energy 0.51 MeV each. The process is called annihilation. The annihilation-generated gamma photons interact in the medium either by photo-electric or Compton process. The process of pair production is in a way the reverse of bremsstrahlung process, both taking place in the field of the atomic nucleus. Out of the three processes of X and gamma ray interaction, it is the Compton effect which predominates in most of the materials in the photon energy range of our interest. The photoelectric effect becomes comparable only at low energies, say below 30 keV in the case of low Z elements such as Li, Be, and B. For medium Z elements of importance in TL such as Al, Ca, Mg, Si, etc., its contribution is significant below 100–200 keV. Pair production makes significant contribution only above about 2–3 MeV.

Above summary shows that the interactions of X or gamma ray photons produce energetic electrons, which carry the energy of the incident photon in the form of their kinetic energy. These fast moving electrons liberate a large number of bound electrons from atoms of the target medium by exerting electrostatic force on them as discussed above. The most loosely bound electrons have the maximum chance of getting freed. In other words, the electrons are liberated from the valence band of a solid phosphor, since it is this energy state which has a minimum binding energy. The energy deposition by X and gamma ray photons in matter is thus a two step process. First, an energetic electron is produced. This electron is indistinguishable from any other incident electron like a beta ray or cathode ray. In the second step, this electron interacts in the matter exactly as described earlier in the case of beta particle. The TL induction rate per unit dose therefore is similar for beta and X/gamma rays of similar linear energy transfer (LET).

2.4.4 Neutrons

Neutrons are electrically neutral particles. They have mass of nearly 1 atomic mass unit like a proton. Being neutral they do not interact with atomic electrons and hence do not produce ionization directly. They undergo two types of reactions with the atomic nuclei, namely: (a) capture, in which the target nucleus absorbs the neutron. As a result of the neutron capture, another type of radiation leaves the nucleus which causes ionization. (b) elastic scattering in which the neutron collides with the nucleus of an atom and transfers some of its energy to it. The capture reactions are predominant for slow (or thermal) neutrons. There are four main types of slow neutron capture reactions. These are (n, γ), (n, α), (n, p), and $(n, \text{fission})$ reactions. In (n, γ) reaction, the compound nucleus after the capture of the neutron is in an excited state. Emission of energy in the form of one or more gamma photons brings the compound nucleus to its ground state. Some examples of this reaction are

$$
\begin{aligned}
{}^{1}_{1}\mathrm{H} + {}^{1}_{0}\mathrm{n} &\rightarrow {}^{2}_{1}\mathrm{H} + \gamma \\
{}^{123}_{48}\mathrm{Cd} + {}^{1}_{0}\mathrm{n} &\rightarrow {}^{124}_{48}\mathrm{Cd} + \gamma \\
{}^{27}_{13}\mathrm{Al} + {}^{1}_{0}\mathrm{n} &\rightarrow {}^{28}_{13}\mathrm{Al} + \gamma \\
{}^{164}_{66}\mathrm{Dy} + {}^{1}_{0}\mathrm{n} &\rightarrow {}^{165}_{66}\mathrm{Dy} + \gamma
\end{aligned}
$$

The gamma photon energy from the first reaction is 2.2 MeV. For all other nuclei, the total gamma energy released is 6–8 MeV, which is equal to the binding energy per nucleon in the compound nucleus. This energy, however, is emitted in the form of many photons of lower energy. These reactions are relevant in TL dosimetry of neutrons. The (n, α) and (n, p) reactions are not so common as (n, γ). Following two reactions are important from the view point of application of TL in neutron dosimetry.

${}^{6}_{3}\mathrm{Li} + {}^{1}_{0}\mathrm{n} \rightarrow {}^{3}_{1}\mathrm{H} + \alpha$ (reaction cross-section $\sigma = 950$ b, kinetic energy of $\alpha = 2.05$ MeV and that of ${}^{3}_{1}\mathrm{H} = 2.73$ MeV, ${}^{3}_{1}\mathrm{H}$ is stripped of the electron)

$$
{}^{10}_{5}\mathrm{B} + {}^{1}_{0}\mathrm{n} \rightarrow {}^{7}_{3}\mathbf{Li} + \alpha (\sigma = 3840\ \mathrm{b},\ \text{kinetic energy of}\ \alpha = \mathbf{2.32}\ \mathrm{Mev})
$$

${}^{6}_{3}\mathrm{Li}$ is used in the well-known TLD-600 phosphor which is used for thermal neutron dosimetry.

An important (n, p) reaction is:

$$
{}^{14}_{7}\mathrm{N} + {}^{1}_{0}\mathrm{n} \rightarrow {}^{14}_{6}\mathrm{C} + {}^{1}_{1}\mathrm{P}
$$

${}^{14}_{6}\mathrm{C}$ is formed in the atmospheric air by the reaction of cosmic ray neutrons with nitrogen. It forms the basis of the dating of organic remains in archaeology.

One of the most important reactions of thermal neutrons is fission of uranium-235 atoms:

Table 2.2 Number of collisions a 2 MeV neutron suffers in different materials before thermalization [14]

Material	No. of collisions
Hydrogen	18
Water	19
Lithium	67
Beryllium	86
Carbon	114
Oxygen	150
Fluorine	177

$$^{235}_{92}\mathrm{U} + {}^{1}_{0}\mathrm{n} \rightarrow \mathrm{x} + \mathrm{y} + 2^{1}_{0}\mathrm{n}$$

This reaction is the basis of the nuclear chain reaction. X and Y are the fission product nuclides. Most of these are radioactive elements emitting beta and gamma radiations.

Elastic scattering of neutrons—it is the most common reaction in the case of fast neutrons. The process is like the collision of billiard balls. The energy ΔE imparted by an incident fast neutron of energy E to the target nucleus of mass A is given by the following equation.

$$\Delta E/E = 2A(1-\mathrm{Cos}\,\theta)/(A+1)^2 \quad (2.4)$$

where θ is the scattering angle of the neutron in the center of mass system. It may be seen that the energy transfer increases as θ increases from 0 to 180°. Again the transfer is much more to the lighter nuclei than to the heavier ones. Obviously maximum energy transfer takes place when the target nucleus is that of hydrogen. In a bulk medium fast neutrons get thermalized by multiple collisions with the atoms of the medium. Estimate of the number of collisions that a neutron of energy 2 MeV makes before getting thermalized is given in Table 2.2 for different materials [14]. Fast neutron collisions have twofold importance in TL induction: (1) the nuclei which suffer collisions with neutrons get displaced into an interstitial position with the result that a vacancy is left behind in the lattice. It has been estimated that the number of vacancies produced by a fast neutron of energy up to 5 MeV runs in couple of hundreds in materials with $Z \leq 30$ [15]. These vacancies may function as traps for electrons or holes depending on whether the vacancy is of negative or positive ion. (2) when a hydrogenous material is kept in contact with a TL phosphor and exposed to fast neutrons, energetic recoil protons produced in abundance as a result of elastic scattering in the former cause ionization and induce TL in the phosphor. Fast neutron dose meters have been made based on this technique. (see for example, Sunta et al. [16] and Rzysky et al. [17]).

2.5 Dose Dependence of the Filling of Traps

As discussed earlier, the TL excitation process consists essentially of filling of the traps. The TL response of a given sample to a known dose depends on the number of the traps filled by the given dose. A simple way to estimate the number of the filled traps is to assume the filling rate to be directly proportional to the dose D and also directly proportional to the vacancies in the traps [18]. If the total number of trapping sites is N out of which n are already filled up at any time, the incremental filling dn by a small incremental dose dD may be given by

$$\mathrm{d}n = \alpha(N - n)\ \mathrm{d}D \tag{2.5}$$

where α is the fraction of vacant traps which get filled up per unit dose. Using the initial conditions that $n = 0$ when $D = 0$, the solution of (2.5) is

$$n = N\left[1 - \exp\left(-\alpha D\right)\right] \tag{2.6}$$

This equation shows that the dose at which all of the traps get filled up depends solely on α. It will be shown in Chap. 6 that the value of α may be determined from the growth curve of TL intensity versus dose, provided however, that the given growth curve fits into the above equation. Another way to monitor the trap filling is to measure the optical density (OA) or ESR signals from the traps responsible for the TL glow peak. This requires establishing a correlation between the given TL glow peak and an OA band or ESR signal of the sample. Such correlations seem to have been established for the prominent glow peaks of LiF:Mg,Ti (TLD-100) phosphor. The glow peaks appearing at about 200 and 400 °C of this phosphor are thought to be associated with the OA bands at 310 and 225 nm, respectively [18, 19]. Such correlated OA bands or ESR signals may, however not be available for the glow peaks of every sample.

An alternative to the above analytical approach to describe the trap-filling process is to use a set of rate equations, which represent the flow of charges between the valence band, the conduction band, and the electron traps and the hole centers following ionization by irradiation. Chen et al. [20] used this approach. The rate equations have to be solved numerically to find the number of filled electrons traps at the end of the irradiation which implies the dose. The results of the numerical computations depend on the values of the input parameters used in the equations. An important result, which has implication for experimental cases is that the population of the filled traps n, produced by irradiation depends on the electron-hole generation rate R by ionization, which means that the TL output for a given dose would depend on the dose rate as well. The authors also showed that the pattern of this dependence changes with the total generation of electrons and holes which means the total dose. The dose rate dependence of TL could have important consequences in dosimetric application. A summary of experimental data given by the authors, however, shows that most of the phosphors used in dosimetry do not display this effect.

References

1. J.H. Schulman, D.W. Compton, *Color Centers in Solids*, Chapter 7 (Pergamon Press, New York, 1962), pp. 10–11
2. W.B. Fowler (ed.), *Physics of Color Centers*, Chapter 2 (Academic Press, New York, 1968), pp. 53–179
3. J.L. Merz, P.S. Pershan, Phys. Rev. **162**, 217–235 (1967)
4. C.M. Sunta, Radiat. Prot. Dosimetry **8**, 25 (1984)
5. C.M. Sunta, Nucl. Tracks **10**, 47 (1985)
6. J.D. Comins, B.D. Carragher, J. Phys. **41**, 166 (1980)
7. P.D. Townsend, Nucl. Instrum. Methods Phys. Res. **197**, 9 (1982)
8. F. Augullo Lopez, F.L. Lopez, D. Jaque, Cryst. Lattice Defects Amorphous Mater. **9**, 227 (1982)
9. C.M. Sunta, J. Phys. C: Solid State Phys. **3**, 1978 (1970)
10. K.S.V. Nambi, V.N. Bapat, A.K. Ganguly, J Phys. C: Solid State Phys. **7**, 4403 (1974)
11. B. Dhabekar, S. Menon, R. Kumar, T.K. Gundu Rao, B.C. Bhatt, A.R. Lakshmanan, J. Phys. D: Appl. Phys. **38**, 3376 (2005)
12. S. Watanabe, T.K. Gundu Rao, P.S. Page, B.C. Bhatt, J. Lumin. **130**, 2146 (2010)
13. M.J. Aitken, Dose rate evaluation, *Proceedings of Specialist Seminar on Thermoluminescence Dating, Research Laboratory for Archaeology and History of Art*, Oxford, PACT vol. 2, part 1 (Council of Europe, Strasbourg, 1978), p. 18
14. R. Stephenson, *Introduction to Nuclear Engineering* (Mcgraw-Hill, New York, 1954), p. 104
15. P.W. Levy, in *A Brief Survey of Radiation Effects Applicable in Geological Problems*, ed. by D.J. McDougall. Thermoluminescence of Geological Materials (Academic Press, New York, 1968), p. 25
16. C.M. Sunta, K.S.V. Nambi, V.N. Bapat, *Symposium on Neutron Monitoring for Radiation Protection Purposes, International Atomic Energy Agency, Vienna*, IAEA/SM-167/10, 11–15 Dec 1972
17. B. Rzysky, S. Watanabe, C.M. Sunta, *Fifth International Conference on Luminescence Dosimetry, University Sao Paolo, Brazil*, Paper no P-52, 14–17 Feb 1977
18. C.M. Sunta, E. Okuno, J.F. Lima, E.M. Yoshimura, J. Phys. D Appl. Phys. **27**, 2636 (1994)
19. L.V.E. Caldas, M.R. Mayhugh, T.G. Stoebe, J. Appl. Phys. **54**, 3431 (1983)
20. R. Chen, S.W.S. McKeever, S.A. Durrani, Phys. Rev. B **24**, 4931 (1981)

Chapter 3
Thermal Stimulation of Luminescence and Theory of the Glow Curves

Abstract The chapter begins with the explanation of the thermal stimulation process and goes on to describe conventional models of thermoluminescence (TL). The characteristic properties of the glow curves of the Randall-Wilkins, the Garlick-Gibson, the general order (GO), and the mixed order (MO) kinetics models are summarized. While dealing with the GO kinetics model it is shown that the pre-exponential factor (PF) s' and the kinetic order (KO) parameter b are not independent constants as assumed by the proponents of the model, but that the s' value is dependent on the value of b and the total concentration N of the traps. The lacunae in GO kinetics model and the shortcomings of MO model are discussed. The chapter then takes up the physical models. These include the multitrap systems which are the simplified version of the generalized scheme consisting of a host of traps and equally large number of recombination centers (RC). These may plausibly be applicable to the real materials. It is shown by simulations how first order (FO) kinetics glow peaks are produced under a variety of parametric conditions. The results of these simulations are used to answer the question why the KO of the TL glow peaks of real materials is invariably seen to be of FO. It is suggested that thermally disconnected deep traps, which may be stable at high temperatures may contribute to the high concentration of RC (radiative or non-radiative) leading to the FO kinetics. It is logical that defects would exist in a crystalline material until it approaches its melting point. Examples are given of some materials in which deep traps have been detected. Another way by which FO kinetics is produced is the local recombination of the thermally excited charge carriers.

3.1 Energy Levels of Solid Phosphors

The energy states of the electrons in a solid are derived from the free space energy levels of the atoms or ions of which the solid is made. The narrow energy levels of isolated atoms and ions are broadened due to the field of their neighboring ions or

C. M. Sunta, *Unraveling Thermoluminescence*,
Springer Series in Materials Science 202,
DOI: 10.1007/978-81-322-1940-8_3, © Springer India 2015

atoms in the solid. The energy levels of the solids thus are broad bands [1]. The uppermost band which is filled with electrons is called the valence band *V*. In alkali or alkali earth halides, this band may be assumed to be filled with the outermost filled *p* state electrons of the halide ions. These electrons can move from one ion to other in the band without changing their total energy state. The next higher energy band is generally empty. This corresponds to the state of an electron free from an ion. It is called the conduction band *C*. In addition to these bands which apply to the whole solid, there are localized electron states which are produced by impurity atoms or electrons or holes in defect centers as already described in Sect. 2.2. These states are comparatively narrow levels and they occur between *V* and *C* energy bands of the pure sample. It is these states that get filled up during irradiation which we termed as induction of thermoluminescence (TL) in the previous chapter. Release of electrons from these shallow bound states (called traps T) by heating is called the thermal stimulation. The recombination of the released electrons at a luminescence center ultimately produces the light which we call TL.

3.2 The Thermal Stimulation Process

The electron in a trap is analogous to a ball fallen into a well under the field of gravity (potential well). Depending on its structure (refer Sects. 2.1 and 2.2) the trap depth varies from one type of traps to another. The trap depth is essentially the binding energy of the charge carrier (electron or hole) to the defect structure in the crystalline lattice. Glow peaks appear at different temperatures due to the emptying of different types of traps. The glow curve structure of the various phosphors is quite complex, each having a number of glow peaks, some resolved and some unresolved. We consider in the following the case of an isolated single glow peak. It then implies that we are dealing with the traps of only one type.

3.2.1 Energy Distribution of Electrons in the Traps

To begin with let us consider the free particles or molecules of a gas in an enclosed volume. These particles are in constant state of motion at any temperature above absolute zero. Although most of the particles have similar average energy at a given temperature, the statistical variations cause some to have the energy considerably in excess of the average and some very small. Application of the laws of mechanics to the free particle motion shows that the statistical variation of the energy follows the Maxwell–Boltzmann distribution. This distribution however, is not applicable when we are dealing with the electrons or holes bound with a given binding energy to a defect center, the energy distribution of these particles in the bulk medium may be dealt in the following manner.

Let us consider the case of an electron population bound to the traps of a given type. We usually describe it as the population of filled traps. At low temperatures, these electrons are at the ground state of the trap. The ground state itself is at E eV below the bottom of the conduction band. The fraction n/n_0 which has risen above the ground state of the trap at a given temperature T is then governed by the Boltzmann function viz,

$$\frac{n}{n_0} = \exp\left(-\frac{\Delta E}{kT}\right) \tag{3.1}$$

where n_0 is the total number of electrons in this population and n is the number with energy equal to ΔE above the ground state of the trap at a given temperature T. From (3.1), we may find the value of n/n_0 for different values of ΔE at any given temperature. Figure 3.1 shows the fraction n/n_0 of the trap population corresponding to different energies, at three different temperatures. This includes the energy distribution at 475 K which is the approximate peak temperature of a prominent glow peak of the LiF TLD phosphor. What is important for us to know however is the fraction (n/n_0) of the total filled trap population, which would have energy ΔE equal to the trap depth E, since it is this population which is able to exit from the trap to produce TL. One may for example see that if the trap depth is 1 eV, the fraction of the electron population which is capable of escaping from the trap at the above stated temperature is about 10^{-13}. At further lower temperatures, this fraction would be still lower. The question one may ask is: with such small fractions attaining sufficient energy to cross the trap barrier, how does the trap become empty? The answer lies in the pre-exponential factor (PF) s. This factor s is appropriately called the 'attempt to escape frequency' or simply the frequency factor and its value may be only slightly less than the frequency of vibration of the atoms/ions of the medium. Taking the analogy of the potential well model of the trap, it means that the trapped particles which have attained the requisite energy to be able to escape from the trap, are striking at the walls of the well at the top of the potential well at a very fast rate of about 10^{14} s^{-1} (vibrational frequency of the particles). With some getting reflected back, the net frequency of success for escaping may still be taken as about 10^{13} s^{-1}. Now if the population fraction with requisite energy is about 10^{-13}, the net probability p for escape from the traps per second becomes $p = s\exp(-E/kT)$, which means nearly equal to 1, in other words the total escape. The factor, which helps to empty the traps is the rapidness with which the thermal equilibrium is re-established after the energetic electrons have escaped. The time involved is of the same order as the inverse of the vibrational frequency of the particles at the TL recording temperature. With the re-establishment of thermal equilibrium next batch consisting of similar population fraction is ready to escape. This process continues till the traps are emptied totally. At this point it may be appropriate to elaborate further on the term "frequency factor".

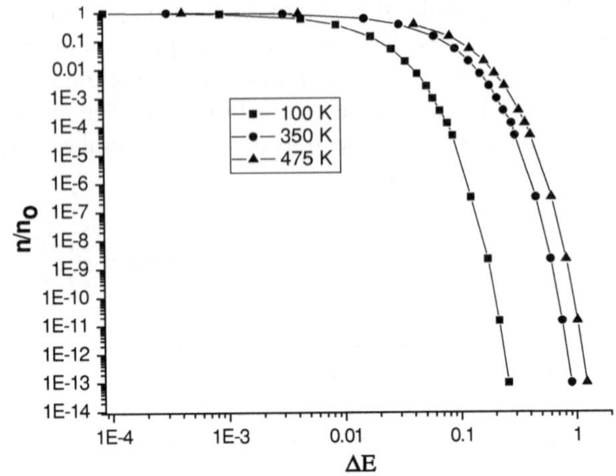

Fig. 3.1 Energy distribution of trapped electron population at temperatures 100, 350 and 475 K. ΔE is the energy of the electrons above the ground state of the trap, n_0 is the total population of the trapped electrons and n is the electron population having energy ΔE at the given temperature

3.2.2 *Frequency Factor*

As discussed above, the untrapping of an electron from a trap of depth E under thermal stimulation involves two steps: first, the electron is elevated to an energy level E above the ground state of the trap by the thermal agitation. The fraction of the trapped electron population that can achieve energy nearly equal to the trap depth E above the ground state of the trap is given by the Boltzmann function, $\exp(-E/kT)$. These electrons as stated above are not yet totally free from the traps. These may again be knocked down into the trap or may transit into the free state (conduction band) in the presence of continuous thermal agitation. The *frequency factor* is the rate of this transition. Any transition rate depends essentially on two factors namely the vacancies in the state into which the transition is to occur and the cross-section for the transition. In the present case, the electron thermal velocity v, also comes into play. According to Bube [2], this may be explained as: The vacancies in the conduction band are equal to the effective density of states of the conduction band, designated as N_c cm^{-3}. Consider the relative motion between the electron in the excited state of the trap and the N_c states per cm^3. In the presence of thermal motion the electron encounters these states $N_c v$ times cm^{-2} s^{-1}. If the transition cross-section of the electron to transit into the conduction band is σ cm^2, the rate of this transition becomes equal to $N_c \sigma v$ (s^{-1}). This gives the value of frequency factor s which, as stated before is also called as attempt to escape frequency. More simply stated the value of s may approximately be equal to the number of times per seconds that a bound electron may absorb energy from the crystal vibrations, multiplied by the probability of it to be thrown into the free state [2]. Thus as stated before the value of s may be a little less than the frequency of the crystal vibrations (Debye frequency). Note here that the cross section σ, according to the principle of detailed balance has the same value as that for the electron to fall from the conduction band into the trap [2]. The parameters N_c, v, and σ have a certain degree of temperature dependence, which render s also temperature dependent. N_c and v are given by

$$N_c = 2\left[\left(\frac{kTm^*}{2\pi h^3}\right)\right]^{\frac{3}{2}} \tag{3.2}$$

$$v = \left(\frac{3kT}{m^*}\right)^{\frac{1}{2}} \tag{3.3}$$

where m^* = effective electron mass and h = Planck's constant divided by 2π. The temperature dependence of σ has been taken as $T^{-3/2}$ by Lewandowki and McKeever [3]. Using these expressions, $s = N_c v\sigma$ is found to have a temperature-dependent factor of $T^{1/2}$. The frequency factor s being a large quantity (for example 10^{13} s^{-1}), a change of temperature, for example, from 300 to 600 K during a glow curve recording would not influence the value of s significantly. One may therefore be justified in ignoring the temperature dependence of s.

3.3 The Thermoluminescence Models

A simple energy level scheme consisting of one type of traps and one type of recombination centers (OTOR) is shown in Fig. 3.2. The traps are assumed to be of electron type and the recombination centers (RC) of hole type. The traps and the RC get filled up by the respective type of charge carriers during irradiation. During heating the electrons are ejected out of the traps and become free to move in the conduction band. During the random motion if the electron comes across an oppositely charged defect center called the RC it may recombine with it and emit luminescence. Alternatively it may fall back into an empty trap (see Fig. 3.3). This is called retrapping. The transport of the electrons from the traps to the RCs to emit luminescence may be described in terms of rate of these three components. These three rates for the OTOR model are given by

$$\text{Excitation rate } R_{\text{ex}} = ns\exp\left(-\frac{E}{kT}\right) \tag{3.4}$$

$$\text{Retrapping rate } R_{\text{ret}} = n_c A_n (N - n) \tag{3.5}$$

$$\text{Recombination rate } R_{\text{rec}} = n_c A_h h \tag{3.6}$$

where N and n are the total and the filled concentration of the thermally active traps, n_c is the concentration of the charge carriers in the free state, h is the concentration of the available RC, E is the activation energy of the trap, which is also called the trap depth, T is the sample temperature, k is the Boltzmann

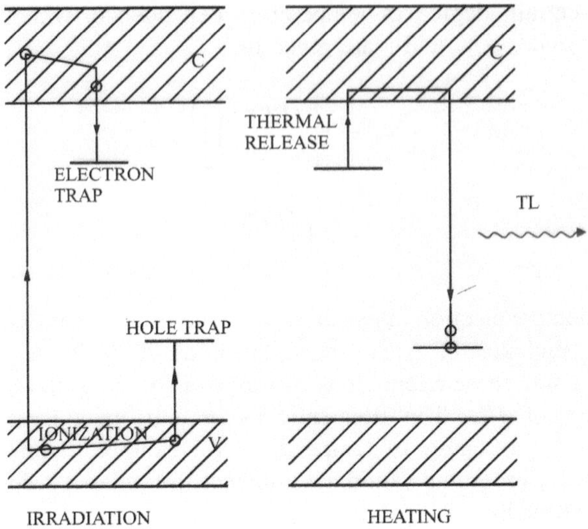

Fig. 3.2 One trap one recombination center (OTOR) model of TL showing the path of the electron transport during irradiation and heating. In the heating phase the released electron is shown to recombine with a hole to emit luminescence, as envisaged in the first order kinetics model of Randall and Wilkins. In this model the probability for the released electron to return to a vacant trap is supposed to be negligible

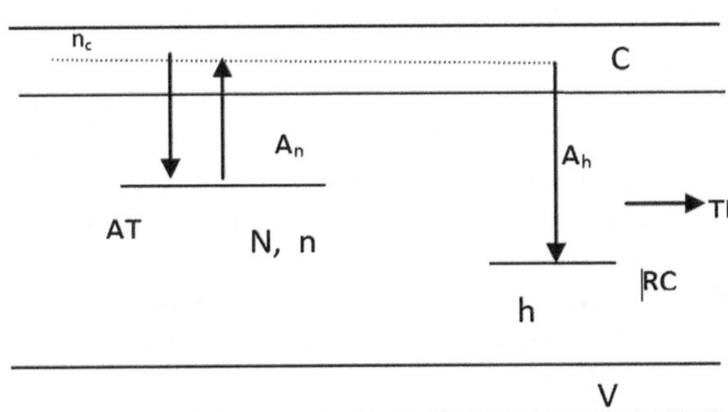

Fig. 3.3 Energy level diagram for Garlick-Gibson model. This model also is based on the OTOR model. In this model the released electron during heating has two options for relaxation namely that it may either recombine to emit luminescence or it may be retrapped by a vacant trap. The probability constants for recombination and retrapping are assumed to be equal ($A_h = A_n = A$)

constant and A_n and A_h, respectively are the retrapping and the recombination coefficients. The values of these coefficients depend on the electron capture cross sections σ_n and σ_h of the traps and the RC, respectively: ($A_n = \sigma_n \upsilon$ and $A_h = \sigma_h \upsilon$ where υ is the free electron velocity in the conduction band). The value of h depends on the physical scheme used in formulating any particular model of TL. For the models which are based on the OTOR scheme h is equal to n.

Essentially the value of h is the sum of all the filled traps in the sample, which means the filled active traps n and any deeper level traps which are thermally not affected, at the given temperature T. The equality between the value of h and sum of all the filled traps ensures the overall charge neutrality of the sample. It is assumed that all the excited charge carriers relax from their free state instantly either into the RC or alternatively into the empty traps, with no significant number of these being left behind in the free state (conduction band when the charge carriers are electrons). This is called the quasi-equilibrium (QE) condition and is discussed in detail in Chap. 5. We may see that the expressions for R_{ex} and R_{ret} would be the same for the various models of TL. The expression for R_{rec} alone would change with the model used. Considering the relaxation probabilities R_{ret} and R_{rec}, the fraction F of the excited carriers which produces luminescence during heating is given by

$$F = \frac{R_{rec}}{(R_{rec} + R_{ret})} \tag{3.7}$$

Depending on the values of the parameters in (3.4)–(3.6), the value of F would change. If there are additional routes of relaxation, for example nonradiative recapture in deeper level traps, these would add to the denominator in (3.7). The expression for F would thus changes according to the applicable physical model (refer to Sect. 3.5).

3.3.1 Randall and Wilkins Model

Randall and Wilkins (RW) [4] were the first to suggest a theoretical model for the TL emission. They assumed that the retrapping may be negligible ($R_{ret} = 0$) and therefore according to (3.7), we have $F = 1$. This means that the TL emission intensity I is directly proportional to R_{ex}.

$$I = c R_{ex} = -c\frac{\mathrm{d}n}{\mathrm{d}t} = cns\exp\left(-\frac{E}{kT}\right) \tag{3.8}$$

where $T = T_0 + \beta t$ and where c is a constant representing the optical efficiency factor relating the luminescence output to the electron release rate and the measuring instrument's efficiency to collect the light. We may take $c = 1$, since it makes no difference to the characteristics like the shape of the glow curve and its decay pattern. It influences only the intensity.

Rearranging (3.8) we have

$$\frac{\mathrm{d}n}{n} = -s\exp\left(-\frac{E}{kT}\right)\mathrm{d}t \tag{3.9}$$

Assume that the sample is warmed up at a linear rate of heating, so that $dT/dt = \beta\ \mathrm{Ks}^{-1}$. This equation may then be written as

$$\frac{dn}{n} = -\left(\frac{s}{\beta}\right)\exp\left(-\frac{E}{kT}\right) \tag{3.10}$$

Integrating this equation we get the value of n at any temperature T during the heating viz

$$n = n_0 \exp\left(-\int_{T_0}^{T}\left(\frac{s}{\beta}\right)\exp\left(-\frac{E}{kT'}\right)dT'\right) \tag{3.11}$$

where n_0 is the initial number of trapped charges and T_0 is the temperature at the beginning of the heating run. Substituting this expression for n in (3.8) we get the expression for TL intensity, as a function of temperature T viz

$$I(T) = n_0 s \exp\left(-\frac{E}{kT}\right)\exp\left[-\int_{T_0}^{T}\left(\frac{s}{\beta}\right)\exp\left(-\frac{E}{kT'}\right)dT'\right] \tag{3.12}$$

This is the expression of the glow curve. In Fig. 3.4 are shown the profiles of change in the filled trap population n, the thermal excitation probability p and the TL intensity I as the as a function of temperature T when the sample goes through the heating cycle to read the glow curve. The initial part of the glow curve rises exponentially. In this part, the change in n is not perceptible. On the other hand the probability of thermal excitation p given by $s\exp(-E/kT)$ rises exponentially. The value of n may be considered constant at n_0 in this part. This property, as we will see later, is useful to determine the value of trap depth E using the initial rise method. When the number of trapped charges n, is appreciably diminished, the TL intensity curve ceases to rise in the exponential fashion. It goes through a maximum before falling and ultimately falls to zero when all the traps are emptied. The intensity I at any temperature T in the glow curve is equal to the product of the values of n and p at that temperature. The intensity I being directly proportional to n (see (3.8)), this model of TL comes under the category of first order (FO) kinetics. The characteristics of the glow curves of this model are:

(1) The glow curve is asymmetric, rising comparatively slowly and falling somewhat sharply (see the glow curve in Fig. 3.4).
(2) The glow peak shifts to higher T as E increases. Higher value of E means stronger binding of the trapped charge. Hence, it is understandable that a

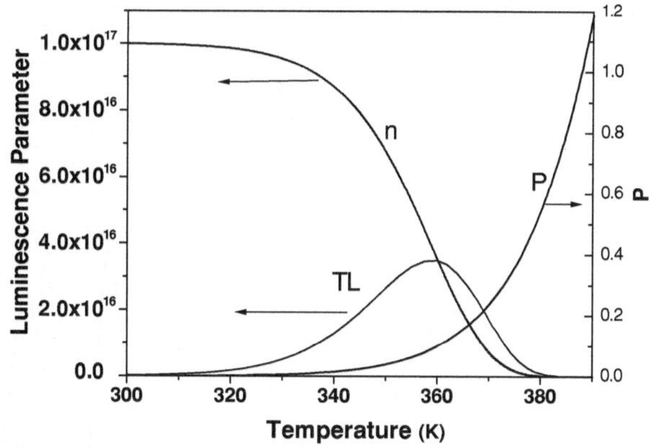

Fig. 3.4 Profiles of trapped electron population n, probability p of excitation and the glow curve as a function of T during heating. Intensity values plotted in the glow curve are 10 times of the actual computed values for better visibility. Calculations are carried out using the Randall-Wilkins model. Input parameters used are $E = 1$ eV, $s = 10^{13}$ s^{-1}, $\beta = 1$ K s^{-1}. Initial value of n is 10^{17} cm^{-3}

higher temperature is needed to release it. Figure 3.5a shows the glow peaks for different E, for given values of s and β. It is seen that with increase in the value of E the peak height is reduced, but the peak half width ω, is increased, thus maintaining the area constant for a given value of n_0.

(3) The temperature of maximum intensity, T_m increases almost linearly with E as seen in Fig. 3.5b.

(4) For given values of E and β, the glow peak shifts to lower T as s is increased (see Fig. 3.6). Thus, E and s have opposite effects on T_m. This is simple to understand since higher s means faster escape of the trapped change from the excited state of the trap. Higher s leads also to increase in peak height and to reduction in the peak width ω.

(5) For given values of E and s, the peak temperature T_m increases as heating rate β is increased. In Fig. 3.7 are shown the glow curves for two different heating rates, plotted in three different ways, namely, (a) intensity I versus time t. When β is increased the glow peak becomes narrower and taller, with area being conserved, (b) intensity I versus temperature T and (c) normalized intensity (I/β) versus temperature T. When the value of I is plotted against T, the peak recorded with higher β increases in height as well as in area. However, when the height is normalized with β the area gets conserved as seen in (c). Kumar et al. [5] have studied the effect of heating rate on the theoretically simulated glow curves as well as on a sample of Al_2O_3:C. The experimental intensity values for different β agree with the theory if the luminescence of the sample does not suffer thermal quenching at higher heating rates.

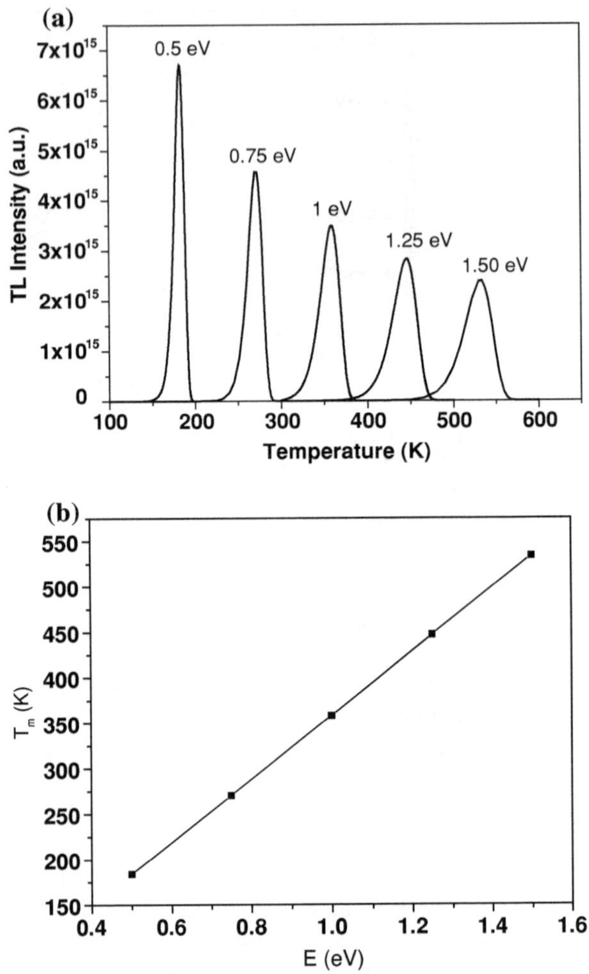

Fig. 3.5 Change in temperature T_m of glow peak maximum with change in E. **a** Glow curves for $E = 0.5$, 0.75, 1, 1.25 and 1.5 eV, **b** T_m versus E. The values of s and β are same as in Fig. 3.4

(6) The glow peak characteristics, namely the peak temperature T_m and the peak shape remain unchanged when the initial filled concentration n_0 of the traps is changed. This means that the peak characteristics are independent of radiation dose given to the sample. However, the intensity (the area as well as the height of the peak) increases in direct proportion to n_0 for given E, s and β. This means that the intensity is directly proportional to dose given to a sample, assuming that the trap filling is directly proportional to the dose. These characteristics are unique to FO kinetics and are of prime importance in the application of TL in radiation dosimetry as well as in kinetics analysis of the glow curves.

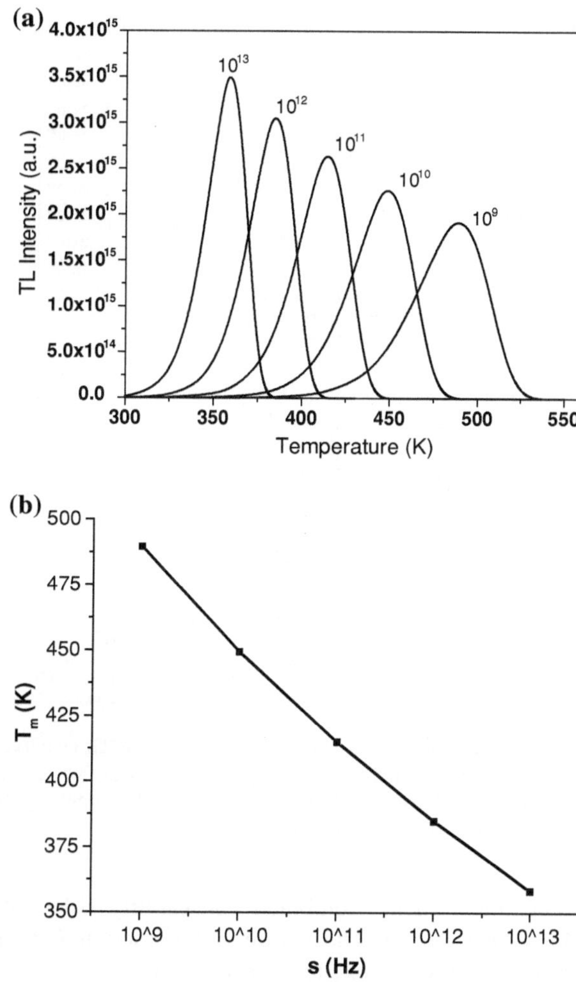

Fig. 3.6 Change in T_m with change in s.**a** Glow curves for $s = 10^{13}$, 10^{12}, 10^{11}, 10^{10} and 10^9 s^{-1}, **b**T_m versus s. Value of E and β are same as in Fig. 3.4

3.3.2 Garlick and Gibson Model

Garlick and Gibson (GG) [6] modified the model of RW [4] for the TL intensity using the same OTOR model. They assume that once an electron is thrown out from the trap into the conduction band, it may either recombine with a RC to produce luminescence or it may be retrapped by any of the vacant traps (see Fig. 3.3). This is in contrast to the RW model in which retrapping is ignored. If we assume that the probability coefficients for recombination and retrapping are A_h and A_n respectively, the probabilities for any excited carrier for recombination and retrapping would respectively be $A_h n$ and A_n $(N - n)$, where N is the total number of the traps and n is the number of available RC at any time. In the OTOR model n is also equal to the number of filled traps, so that the charge balance is

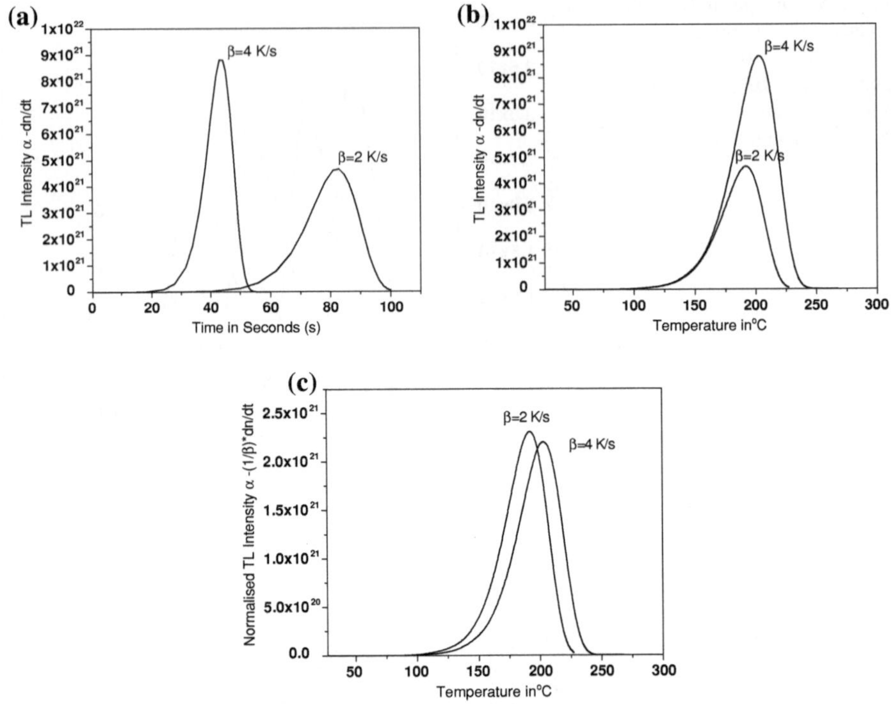

Fig. 3.7 Change in glow curves with change in β. Glow curves recorded with heating rate $\beta = 2$ and 4 K^{-1} plotted in three different ways; **a** intensity versus time, **b** intensity versus temperature, **c** normalized intensity (I/β) versus temperature. The input parameters are $E = 1.1$ eV, $s = 10^{11}\ s^{-1}$ and $n_0 = 10^{23}\ m^{-3}$ (from Kumar et al. [6])

maintained. The combined probability of both the transitions thus is $A_h n + A_n(N - n)$. The recombining fraction F of this combined probability of transitions for any excited carrier, then is

$$F = \frac{A_h n}{A_h n + A_n(N - n)} \tag{3.13}$$

GG assume that the excited charge carrier has no particular preference for recombination or retrapping which means $A_h = A_n$. Thus we get $F = n/N$. In contrast to this, the F value in RW model is equal to 1. We now see that the intensity $I(T)$ of luminescence given by (3.8) would be modified by a factor equal to n/N. Thus, in GG model

$$I(T) = -\frac{dn}{dt} = \left(\frac{n}{N}\right) ns \exp\left(-\frac{E}{kT}\right) = \left(\frac{n^2}{N}\right) s \exp\left(-\frac{E}{kT}\right) \tag{3.14}$$

The intensity I being proportional to n^2, the TL in this model is called by the name as TL of second order (SO) kinetics. Assuming as usual $dt = dT/\beta$, integration of (3.14) yields the value of n at any temperature T. Multiplying that value of n by the escape probability p (= s exp ($-E/kT$)), gives the equation for glow curve viz

$$I(T) = n_0^2 \frac{s}{N} \exp\left(-\frac{E}{kT}\right) \left[1 + \frac{n_0 s}{\beta N} \int_{T_0}^{T} \exp\left(-\frac{E}{kT'}\right) dT'\right]^{-2} \tag{3.15}$$

It is worthwhile to mention here that SO kinetics is obtained in OTOR model even if $A_h \neq A_n$. The required condition for that is $n \ll N$ (low dose sample), so that A_nN becomes much greater than A_hn. Thus $F = A_hn/A_nN$ and in (3.14), n/N gets replaced by A_hn/A_nN and in (3.15), s/N gets replaced by $s\ A_h/A_nN$. It is to be remembered that in the case of $A_h \neq A_n$, the SO kinetics is obeyed only in low dose samples ($n \ll N$), whereas the GG model of SO kinetics is valid at all doses. Figure 3.8 shows the glow curves computed from (3.15) for different values of n_0/N. The characteristics of these glow curves may be described as

(1) The fall of the glow curve is slightly slower than its rise. This is in contrast to the FO kinetics case in which the fall is faster than the rise. The average value of the shape factor for different E and s has been found to be 0.52 in contrast to 0.42 in the case of FO kinetics Chen [7].
(2) For a given set of parameters E, s, β and n_0/N the temperature of maximum intensity (T_m) is on the higher side as compared to that of the FO, except when $n_0 = N$ (traps filled to saturation), in which case the T_m for the FO and the SO kinetics is nearly same.
(3) The glow peaks shift to higher temperature and become progressively wider in shape as the value of n_0/N is reduced, i.e. smaller is the dose, higher the T_m and wider are the peaks. This is in contrast to the FO kinetics case in which T_m and shape factor are independent of n_0/N.

The other properties, namely the dependence of T_m on E, s and β as expected, are similar to that of the FO kinetics case.

At this point it is worthwhile to elaborate on the term kinetic order (KO) which we would be using frequently while discussing different models to describe the TL emission process.

3.3.2.1 Kinetic Order

The term KO in TL is borrowed from chemical kinetics. When the rate of a chemical reaction is described by the change in the concentration of only one of the reacting chemicals it is called mono molecular kinetics or FO kinetics. When it is described by the change in the concentration of both the reactants, it is called bi-molecular kinetics or SO kinetics. In TL, FO kinetics means that the TL

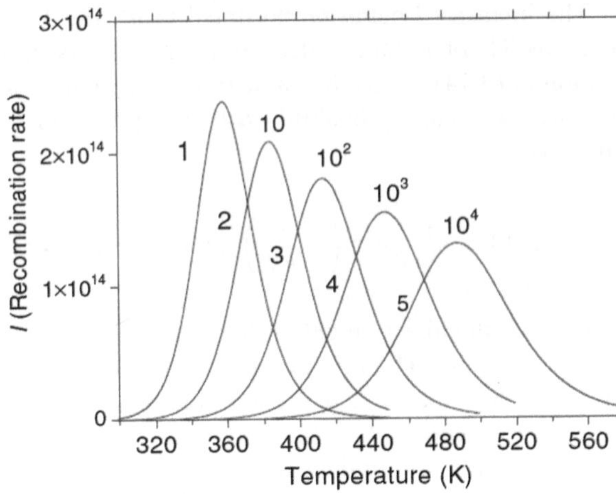

Fig. 3.8 Glow curves of second order kinetics (GG model) for different n_0/N. The parameter values are $E = 1$ eV, $s = 10^{12}$ s^{-1}, $\beta = 1$ Ks^{-1}. Glow curves 1–5 are for $n_0/N = 1, 10^{-1}, 10^{-2}, 10^{-3}$ and 10^{-4} respectively. The intensity values on curves 2, 3, 4, and 5 are multiplied by factors 10, 10^2, 10^3 and 10^4 respectively [22, Fig. 3] © IOP publishing Ltd, reproduced by permission, all rights reserved

intensity I at any temperature during heating depends only on the concentration n of the active traps at that temperature. If the values of I are dependent also on the concentration h of the recombination centers (RC), it becomes a case of non-FO kinetics. In the particular case of OTOR model when $A_h = A_n$. and the concentration h of RC, is equal to the concentration n of the filled active traps ($h = n$), (so that the over-all charge neutrality is maintained), it becomes a case of SO kinetics. Because under the condition $A_h = A_n$, the reaction rate between the released charges from the traps and the RC becomes equal to n^2 which means we have KO equal to 2 (refer Sect. 3.3.2). However when $A_h \neq A_n$ we would not get KO = 2. We would rather get a KO value between 1 and 2 (see Sect. 3.3.4). Thus we see that in contrast to chemical kinetics in which one describes the chemical reactions in terms of either FO or SO kinetics, in TL, in addition to FO and SO (KO = 1 and KO = 2), all values of KO between 1 and 2 also are possible. Such cases which do not fit in FO and SO kinetics, are categorized as general order (GO) kinetics. This is discussed in the following.

3.3.3 General Order Kinetics

May and Partridge (MP) [8] proposed the expression for TL intensity I as

$$I = -\frac{\mathrm{d}n}{\mathrm{d}t} = s' n^b \exp\left(-\frac{E}{kT}\right) \tag{3.16}$$

where s' and b are the empirical constants called PF and the KO, respectively. Other quantities have the same meaning as defined before. The intention of the proponents of this expression had been to provide a general expression for TL emission which would satisfy not only the FO and SO kinetics expressions of their predecessors [4, 6], when $b = 1$ and $b = 2$ respectively, but would also include all other plausible values of b including its non-integral values between 1 and 2 or even outside this range. It is therefore called the GO kinetics model. Although this expression was proposed by its authors by using the same OTOR model as shown for the GG model, the general practice now is to apply it to any experimental glow peak particularly to do the kinetic analysis. Thus implicitly, it is assumed to encompass all plausible physical schemes that may be applicable to the glow peaks. This expression, when solved as a function of T gives the equation for the TL glow peak, viz.

$$I = s' n_0^b \exp\left(-\frac{E}{kT}\right)\left[\frac{(b-1)s' n_0^{b-1}}{\beta}\int_{T_0}^{T}\exp\left(-\frac{E}{kT'}\right)\mathrm{d}T' + 1\right]^{-\frac{b}{b-1}} \tag{3.17}$$

Chen [7] made a simplification in this equation by assuming $s' n_0^{b-1} = s$. Since the function $s' n_0^{b-1}$ has the dimension s^{-1} like in frequency factor s, Chen assumed its role to be similar to that of frequency factor s in the FO case. However, to avoid confusing, it with the frequency factor, later workers have designated it as s''. With the above simplification, (3.17) becomes

$$I = s'' n_0 \exp\left(-\frac{E}{kT}\right)\left[\frac{(b-1)s''}{\beta}\int_{T_0}^{T}\exp\left(-\frac{E}{kT'}\right)\mathrm{d}T' + 1\right]^{-\frac{b}{b-1}} \tag{3.18}$$

The condition for maximum of TL intensity (T_m) in the glow curve is found by equating the derivative (dI/dT) of (3.18) to zero. The condition turns out as

$$\left[(b-1)\frac{s''}{\beta}\right]\int_{T_0}^{T_m}\exp\left(-\frac{E}{kT}\right)\mathrm{d}T + 1 = \left(\frac{s'' b k T_m^2}{\beta E}\right)\exp\left(-\frac{E}{kT_m}\right) \tag{3.19}$$

Chen [7] computed the temperature T_m at the maximum of the glow peak numerically for given values of E, s'', b and β using this equation. Apart from the effect of E and s'' on T_m which is similar to that on FO and SO kinetics, an important result of this work is that the peak shape factor μ given by δ/ω (where ω is the full width at half the peak maximum and δ is the fall side half width of the glow peak) increases monotonically with increase in the value of b and is found to be independent of E and s'' within ±7 %. The graphical relation between μ and b for various values of E and s is given in Fig. 3.9. The average value of δ/ω versus b relation has been routinely used by the TL workers to find the KO of the glow peaks of the experimentally measured glow curves of various TL materials. Another result in the simulations of Chen [7] shows that the T_m is nearly independent of b (see Table 3.2). Chen also showed analytically by simplifying (3.19)

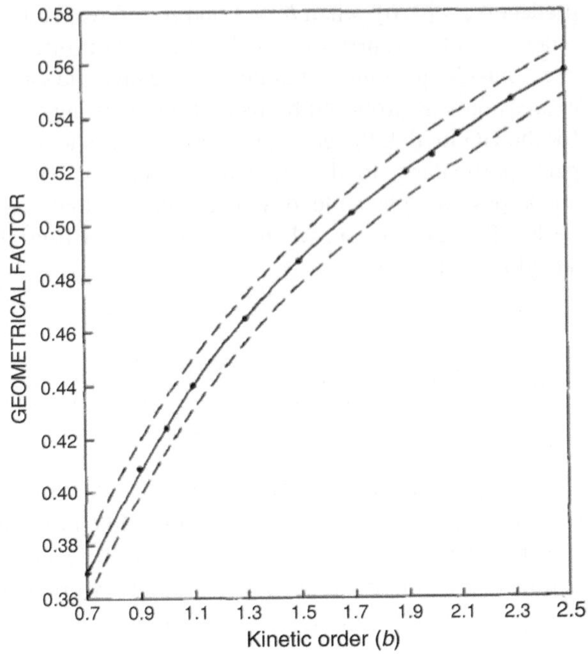

Fig. 3.9 Shape factor μ of a glow curve versus applicable kinetics order [7, Fig. 1]. Reproduced by permission of ECS—The Electrochemical Society

that changes in b do not produce significant change in T_m. This result however, is anomalous, since even by the simple physical understanding of GO kinetics T_m should increase with increase in b. This is because an increased value of b means greater degree of retrapping which is bound to raise the T_m value. This anomaly has resulted as a consequence of assuming $s'n_0^{b-1} = s''$ as a constant. This will become further clear from the following sections wherein we discuss about the inter dependence of s'' and b. Examples of T_m and μ values for different b for selected E and s''combinations are listed below from Chen [7].

b	$E = 1.6$ eV $s'' = 10^{13}\ \text{s}^{-1}$	$E = 0.4$ eV $s'' = 10^{13}\ \text{s}^{-1}$	$E = 0.1$ eV $s'' = 10^{13}\ \text{s}^{-1}$	μ (range)
2.5	554.1 K	–	–	0.548
1.9	554.6 K	144.3	37.6	0.509–0.511
1.5	554.9 K	144.4	37.6	0.476–0.477
1.1	555.2 K	144.5	37.6	0.431–0.432

At this point, it is relevant to discuss the meaning of the empirical parameter s' in (3.16). The strangeness of its units as discussed below indicates that s' may not be an independent constant assumed by MP [8].

3.3.3.1 Pre-exponential Factor, s′ in GO Kinetics

If in (3.16) the TL intensity I, is measured in units of $cm^{-3}s^{-1}$, the units of s' turn out to be $cm^{3(b-1)}s^{-1}$. Let alone the strangeness of these units of s', these units indicate that the s' is not an independent constant as assumed by the proponents of expression (3.16), but it is a function of b and also a function of the active trap concentration (cm^{-3}). In order to find out the relation of s' with b, we may look into the TL expression of FO and SO kinetics, since the GO kinetics expression (3.16) should reduce to FO and SO expressions when $b = 1$ and $b = 2$ respectively. When we substitute $b = 1$ in the above-stated strange units of s', the units turn out to be s^{-1}, which are the units of the fundamental quantity s called the frequency factor. Similarly by substituting $b = 2$ the units of s' turn out to be $cm^3\ s^{-1}$. The latter exactly are the units of s' in SO kinetics [6], since $s' = \frac{s}{N}$ or $s' = \frac{sA_h}{NA_n}$ in SO kinetics (see Sect. 3.3.2). This leads us to infer that the PF in GO kinetics may be given by the function

$$s' = \frac{s}{N^{b-1}} \tag{3.20}$$

where s is the frequency factor as defined earlier and N is the total concentration of the active traps. Sunta et al. [9] have found the values of s' for $1 < b < 2$ directly by interpolation. They begin the interpolation by using the arithmetic mean (AM) of the b values and the geometric mean (GM) of the corresponding s' values at $b = 1$ and $b = 2$. Accordingly for $b = 1.5$, the value of s' would be the GM of s and s/N that is $s' = [s \times s/N]^{1/2}$. The interpolation process may then be repeated for the next midpoint values of b, that is, between $b = 1$ and $b = 1.5$ and also between $b = 1.5$ and $b = 2$. The procedure may be further repeated for finer distribution of b and s' pairs. This produces the following pairs of b and s' values:

b	s'
1	S
1.125	$s/N^{0.125}$
1.25	$s/N^{0.25}$
1.5	$s/N^{0.5}$
1.75	$s/N^{0.75}$
1.875	$s/N^{0.87}$
2.0	s/N
2.5	$s/N^{1.5}$

The graphical relation between s' and b from the above interpolation is shown in Fig. 3.10. It may be seen that each of the above values of s' fits into the common relation $s' = \frac{s}{N^{b-1}}$. The same relation was proposed by Rasheedy [10] to balance

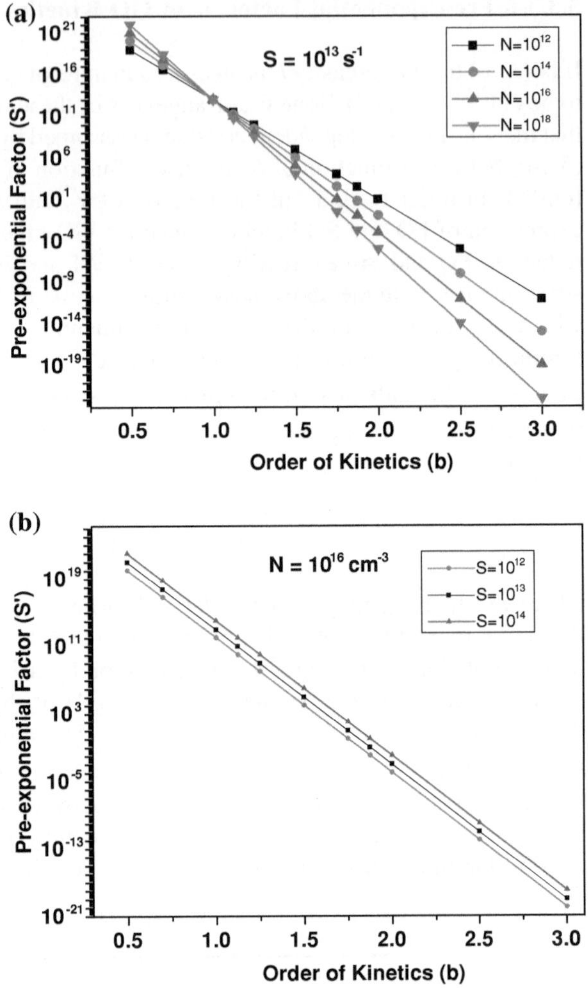

Fig. 3.10 **a** s' versus b plots for $s = 10^{13}$ s^{-1} and $N = 10^{12}$, 10^{14} and 10^{18} cm^{-3}. **b** s′ versus b plots for $N = 10^{16}$ cm^{-3} and $s = 10^{12}$, 10^{13} and 10^{14} s^{-1}

the dimensions in (3.16). If we use $s' = \dfrac{A_h s}{A_n N}$ in place of $\frac{s}{N}$ in SO kinetics, (refer Sect. 3.3.2), the parameter N would get changed to cN (where c is a constant equal to A_n/A_h) in the expression of PF given by (3.20).

We may now return to the treatment of GO kinetics model. Substituting $s' = \frac{s}{N^{b-1}}$, the GO kinetics (3.16) may now be expressed as

$$I = \left(\frac{n}{N}\right)^{b-1} ns \exp\left(-\frac{E}{kT}\right) \tag{3.16a}$$

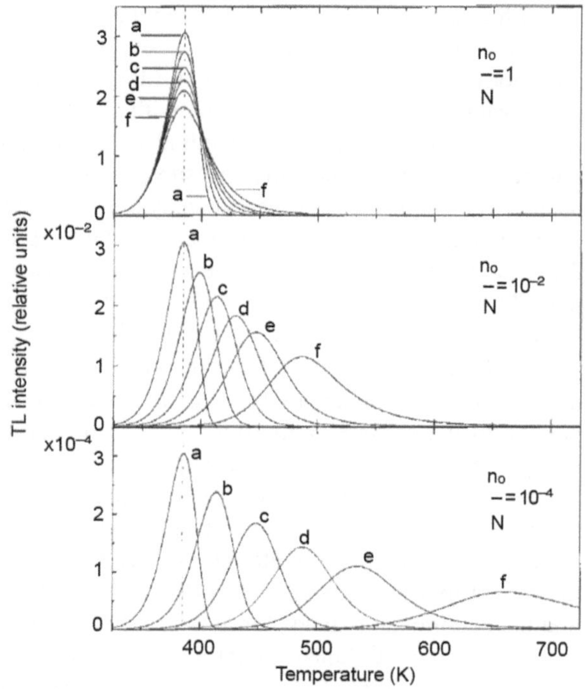

Fig. 3.11 Glow curves of GO kinetics for different values of b and n_0/N. Input parameter values are $E = 1$ eV, $s = 10^{12}$ s^{-1}. Curves in three parts are for $n_0/N = 1$, 10^{-2} and 10^{-4} respectively. The b values for curves a, b, c, d, e and f respectively are 1.1, 1.25, 1.5, 1.75, 2 and 2.5. TL intensity values for $n_0/$N $= 10^{-2}$ and 10^{-4} are multiplied by 10^2 and 10^4 respectively [11, Fig. 2] © Reproduce by permission of Oxford University Press

Integrating it, the equation for the glow curve of GO kinetics turns out as

$$I(T) = s\left(\frac{n_o}{N}\right)^{b-1} n_0 \exp\left(-\frac{E}{kT}\right)\left[\frac{s(b-1)}{\beta}\left(\frac{n_0}{N}\right)^{b-1}\int\limits_{T_0}^{T}\exp\left(-\frac{E}{kT'}\right)dT' + 1\right]^{\frac{-b}{b-1}} \tag{3.21}$$

It may be seen that this equation reduces to that of the SO kinetics when $b = 2$ and to that of the FO kinetics when $b \to 1$. Glow curves derived from this equation for different b and different n_0/N have been given by Sunta et al. [11] and are shown in Fig. 3.11. It may be seen that, (1) the glow curves become progressively wider as the b value increases and (2) glow peaks shift progressively to higher T_m as b increases. This characteristic shift in T_m with change in b did not reveal itself in the calculations of Chen [7] which as stated earlier is because of assuming $s'n_0^{b-1}$ as a constant. However, following Chen various reviews and text books depict the GO kinetics peaks at same T_m for different b to high-light the change in peak shape with change in b (see for example Kirsh [12] and Chen and McKeever [13]). It needs to be stressed that the glow peaks for different b appear at same T_m only under the special condition $n_0/N = 1$ (see Fig. 3.11). The emphasis on the change of peak shape and taking no cognizance of the change in

T_m with b seems to have led the community of TL workers to analyze the glow peaks merely on the basis of the peak shape. Although the peak shape factor δ/ω does not change with trap occupancy (dose), it is important to take note that the glow peaks always shift the position with change in dose (n_0/N) in all the non-FO cases. This provides a simple method to differentiate between FO and non-FO kinetics of the experimentally measured glow curves. The experimentalists may take advantage of this property in analyzing the glow curves and may not need to go through the tedious process of using the GO expression once it is found that the T_m remains unchanged when dose is changed. However, the unique property of non-FO ($b \neq 1$) peaks that T_m shifts with trap occupancy (dose) is seen to have been invariably ignored in analyzing the glow peaks. It is easy to comprehend why T_m should increase with increase in b and with decrease in $\frac{n_0}{N}$. According to the meaning of KO, higher the value of b, greater would be the retrapping in comparison to the recombination during the glow curve recording. Logically higher retrapping rate would delay the emptying of the trap. This in its turn results in delayed (higher T_m) and slower (wider peak) emission of TL. One may similarly understand why glow peaks shift to higher T_m with decrease in n_0/N. The retrapping to recombination rate ratio at any given temperature in the glow curve is directly proportional to $(N - n)/n$ (the numerator gives the vacancies in the traps and the denominator stands for available RC at any given T) in the OTOR model, from which the GO kinetics expression is derived. Smaller the value of n greater would be the ratio $(N - n)/n$, which means greater retrapping. Thus the effect of decrease in n_0/N would be the same as that of increase in b, i.e. higher T_m and wider glow peaks. This is what is seen in Fig. 3.11. In our experience we have not found any real material in which non-first order glow peaks are observed, eventhough superimposition of some major and minor peaks may produce a shape akin to nonfirst order. The criterion for distinguishing a non-first order peak is the shift in peak position with change in dose (trap occupancy) in the cases of all non-first order kinetics. Many of the authors nevertheless report non-first order in their experimental samples without attempting this test. Takeuchi et al. [14], for example, have reported that in MgO doped with iron, low concentration sample shows SO behavior which with higher concentration has $b = 1.3$.

3.3.4 Comparison of GO Kinetics with the OTOR Physical Model

In contrast to the purely empirical treatment of MP [8], Halperin and Braner [15], described the TL emission in a physically meaningful manner. This approach was further developed by Dussel and Bube [16] and Saunders [17]. In this approach too the OTOR scheme is used. The energy-level scheme of the OTOR model is shown in Fig. 3.2. In OTOR scheme the transport of the untrapped charge carriers during the heating may be described by the following rate equations:

$$\frac{\mathrm{d}n}{\mathrm{d}t} = -ns\exp\left(-\frac{E}{kT}\right) + A_n(N-n)n_c \tag{3.22}$$

$$\frac{\mathrm{d}n_c}{\mathrm{d}t} = ns\exp\left(-\frac{E}{kT}\right) - A_n(N-n)n_c - A_h(n+n_c)n_c \tag{3.23}$$

From (3.22) and (3.23), the TL intensity I may be expressed as,

$$I = A_h(n+n_c)n_c = -\frac{\mathrm{d}n}{\mathrm{d}t} - \frac{\mathrm{d}n_c}{\mathrm{d}t} \tag{3.24}$$

Now if we assume $n_c \ll n$ and $\frac{\mathrm{d}n_c}{\mathrm{d}t} \ll \frac{\mathrm{d}n}{\mathrm{d}t}$, the conditions which are called the QE approximations, we can obtain an analytical expression for the TL intensity I for the OTOR model. Using $n_c \ll n$ in (3.24), we have $I = A_h n n_c$. The expression for n_c may be found from (3.23) by assuming $\frac{\mathrm{d}n_c}{\mathrm{d}t} \approx 0$ and $n_c \ll n$

This yields,

$$n_c = \left[\frac{1}{A_n(N-n)n + A_h n}\right] ns\exp\left(-\frac{E}{kT}\right) \tag{3.25}$$

$$I = \left[\frac{A_h n}{A_n(N-n)n + A_h n}\right] ns\exp\left(-\frac{E}{kT}\right) \tag{3.26}$$

where the function within the square brackets is the recombining fraction F as defined earlier (see (3.8)). Using $\frac{A_n}{A_h} = R$, (3.26) may be put in the form.

$$I = \left[\frac{n}{(1-R)n + NR}\right] ns\exp\left(-\frac{E}{kT}\right) \tag{3.27}$$

Now if $R \ll 1$ and $NR \ll 1$ (3.27) reduces to the RW form (FO kinetics) and if $R = 1$, it reduces to the GG form (SO kinetics). When $0 < R < 1$, it becomes a case of GO kinetics. We may therefore say that $0 < R < 1$ corresponds to $1 < b < 2$, where b is the KO value in GO kinetics. We may thus call expression (3.27) as the physical alternative to the empirical GO kinetics expression (3.16). Sunta et al. [18] have numerically calculated the glow curves using (3.27) which are shown in Fig. 3.12. The input values of the parameters namely E, s, n_0/N and β are same as those used in Fig. 3.11, except that in Fig. 3.12, we have the parameter R in place of b. The properties of the glow peaks of Fig. 3.12 may be summarized as:

(1) The peaks shift to higher temperature as R increases. However, at $n_0/N = 1$, the peaks appear at the same temperature irrespective of the R value, until $R > 1$.

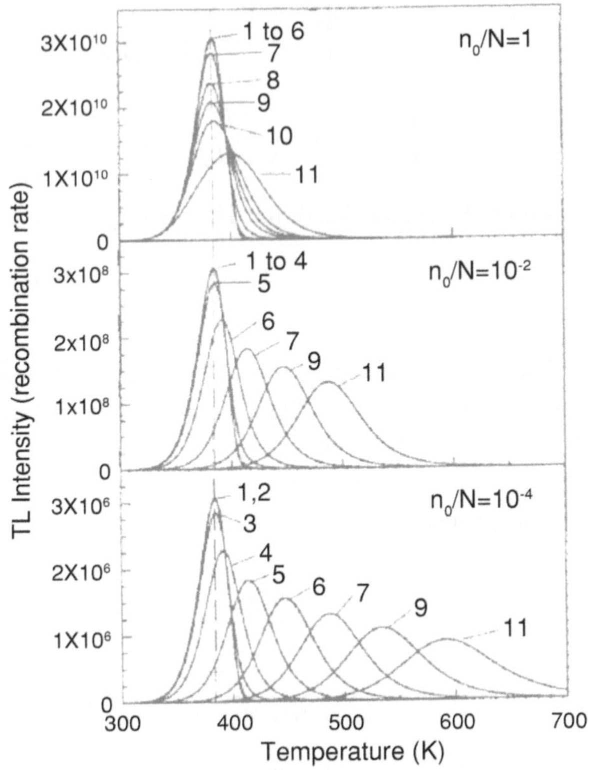

Fig. 3.12 Glow curves for the OTOR scheme calculated numerically using the analytical (3.27). The input values of E, s, β and n_0/N are same as in Fig. 3.11. Glow curves 1–11 are for $R = 0$, 10^{-6}, 10^{-5}, 10^{-4}, 10^{-3}, 10^{-2}, 10^{-1}, 0.5, 1, 2, and 10 respectively [19, Fig. 2] © IOP Publishing Ltd, reproduced by permission, all rights reserved

(2) The glow peaks shift to higher temperature as the trap occupancy n_0/N (dose) decreases.

(3) The width of the glow peaks increases with increase in R and decrease in n_0/N. These characteristics are similar as those of GO kinetics shown in Fig. 3.11. The peak temperature T_m and the peak width ω change with n_0/N and with R exactly in the same fashion as seen in Fig. 3.11 for different n_0/N and different b values. These results show that b and R influence the glow curves in the same way. We may therefore expect a unique relation between b and R values. In order to find this relation, Sunta et al. [18] fitted the glow curves of Fig. 3.12 to the GO kinetics expression. The results show that the relation between R and b depends strongly on the trap occupancy, n_0/N. This means that if the physical OTOR model is applied to an experimental glow curve, the b value of that glow curve would change with the trap occupancy (dose) of the sample. For example at trap occupancies, $n_0/N = 1$, 10^{-2} and 10^{-4}, the best fitted b values for the glow curves of $R = 10^{-3}$ case are 1, 1.3 and 2 respectively (see Fig. 3.13 at $R = 10^{-3}$). This multiplicity of b values for a given R value raises questions about the soundness of the GO kinetics model. Moharil and Kathuria [20] suggested that in place of b one could determine R, which is a physically meaningful quantity. The

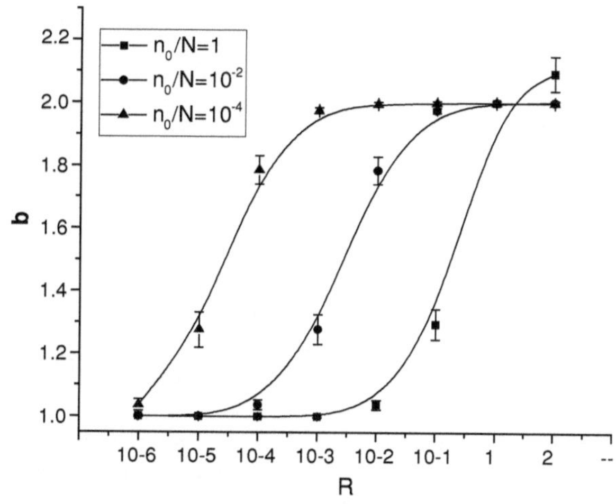

Fig. 3.13 *b* versus *R* curves for three values of n_0/N used in Fig. 3.12. The *b* values plotted are the best fit values for the glow curves of Fig. 3.12 taken from Sunta et al. [9]

dependence of *b* on the system parameters was discussed also by Opanowicz [21] who proposed a kind of effective value of *b* which depends on the system parameters. This is however, of little use as far as application to experimental glow curves is concerned, since the parameters are least likely to be known for a TL phosphor. As of now a *b* value based on the peak symmetry may be considered as the effective value.

3.3.4.1 Lacunae in the Theory of GO Kinetics

In analyzing the experimental glow curves using the GO expression, it is assumed that the GO model encompasses all types of physical models that may apply to the glow curves of various materials. If we apply the GO kinetics equation to the glow peaks of the OTOR physical model we may legitimately equate the TL intensity expression of the OTOR model (see (3.27)) to that of the GO model (see (3.16)) when the values of *E*, *s* and *N* are same in the two expressions. This leads to (See Sunta et al. [22])

$$\frac{N}{(1-R)n+NR}=\frac{s'}{s}n^{b-1} \tag{3.28}$$

Using $s'=\frac{s}{N^{b-1}}$ we get

$$\left(\frac{N}{n}\right)^{b-1}=R\left[\left(\frac{N}{n}\right)-1\right]+1$$

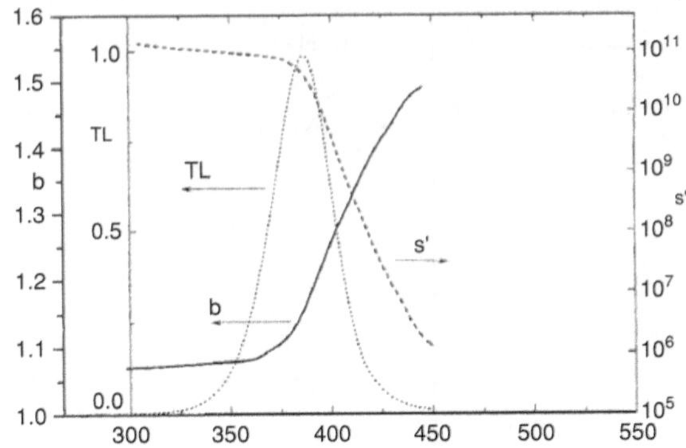

Fig. 3.14 Variation of b and s' with T during heating phase to record the TL glow. (Refer (3.29)). The glow curve is computed using the OTOR model with the parameter values $E = 1$ eV, $s = 10^{12}\ \mathrm{s}^{-1}$, $N = 10^{12}\ \mathrm{cm}^{-3}$, $n_0 = 3.3 * 10^{10}\ \mathrm{cm}^{-3}$, $R = 10^{-2}$ and $\beta = 1\ \mathrm{K\ s}^{-1}$. The shape factor of the glow curve is 0.46 according to which the b value of the glow curve is 1.3 [23, Fig. 8a] © IOP Publishing Ltd, reproduced by permission, all rights reserved

which yields

$$b = \frac{\ln\left\{R\left(\frac{N}{n} - 1\right) + 1\right\}}{\ln\left(\frac{N}{n}\right)} + 1 \tag{3.29}$$

Since n is a temperature dependent variable, it means that b would be varying with T along the glow curve during heating to record the glow curve. The values of n depend also on the initial trap occupancy n_0. This means b becomes dependent also on the initial trap occupancy. This is exactly what is seen in Fig. 3.13. The values of b plotted in Fig. 3.13, are found by the best fit method applied to the glow peaks of Fig. 3.12. These values of b may be considered as some kind of average for the quantity which is varying with temperature during the glow curve reading. An example of how b changes with T along the glow curve is shown in Fig. 3.14. As a consequence of the variability of b, s' also varies with T, since the latter is a function of b (see (3.20)). There are only two special conditions under which (3.29) produces b values which are independent of n. These conditions are $R \to 0$ and $R = 1$ which produce $b = 1$ (FO kinetics) and $b = 2$ (SO Kinetics) respectively. Correspondingly s' values become s and $\frac{s}{N}$, respectively.

The above conclusion that b and s are not constants, but they actually are variable, is based on the application of GO kinetics expression to the OTOR physical model. It may, however, be shown that application of GO kinetics to the glow peaks of any physically plausible model also leads to the same conclusion. In any physical model, the TL intensity I may be expressed as

$$I = F\left[ns\exp\left(-\frac{E}{kT}\right)\right] \tag{3.30}$$

where F is the fraction of the untrapped carriers that recombines to produce TL (recombining fraction). The quantity F is a function of n and includes also other plausible variables depending on the physical model, such as the filled population m in the case of interactive multi-trap system (IMTS) (refer Sect. 3.5). In GO kinetics F is given by

$$F = (s'/s)\,n^{b-1} \tag{3.31}$$

To fit the GO kinetics to a glow peak which is described by the general (3.30), means equating the function F of (3.30) to that of (3.31). This leads to

$$b = n\frac{F'}{F} + 1 \tag{3.32}$$

where F is the recombining fraction of (3.30) and F' is the first derivative of F in n. The value of b thus is clearly seen to depend on n which in its turn is dependent of T during heating. Additionally b would depend also on other variables such as m contained in the function. Here again b can become independent of n and m only under two special conditions, namely $F = c$ and $F = cn$ (where c is a constant), which conditions produce $b = 1$ and $b = 2$ respectively. As a final conclusion, we may state that the GO expression may not faithfully fit into the glow peaks of physical models, unless the results of the best fit produce $b = 1$ or $b = 2$. This means that the GO expression may at best be considered as an approximation. How reliable are the E values of the experimental glow curves of real materials, when determined by fitting these to expressions (3.17) or (3.18) of the GO kinetics is dealt in the next Chapter which deals with the determination of kinetic parameters of glow peaks of real materials.

3.4 Mixed Order Kinetics

The mixed order (MO) kinetics was originally proposed by Visocekas [24], ostensibly to provide a better alternative to the empirical GO Kinetics expression for dealing with TL glow peaks which fit neither in FO nor in SO kinetics. Chen et al. [25] have further elaborated on the model of Visocekas and opined that MO model has merit over GO, because it is derived from a physically meaningful model whereas the GO is only empirical. The scheme of MO kinetics is a simple modification of the OTOR model. It introduces a constant C which adds to the number of RC. The population of RC thus becomes equal to $n + C$. The expression for F which in OTOR scheme is given by the function within squire brackets in (3.26) gets modified for MO model as

$$F = \frac{A_h(n+C)}{A_n(N-n) + A_h(n+C)} \tag{3.33}$$

and the expression for TL intensity I becomes

$$I = \frac{A_h(n+C)}{A_n(N-n) + A_h(n+C)} ns \exp\left(-\frac{E}{kT}\right) \tag{3.34}$$

Now like in the GG model if $A_n = A_h$, (3.34) reduces to

$$I = \frac{s(n+C)}{(N+C)} n \exp\left(-\frac{E}{kT}\right) \tag{3.34a}$$

Alternatively if it is assumed that the trap occupancy is low ($n \ll N$) and additionally the retrapping rate predominates over the recombination rate ($(A_n\,(N-n) \gg A_h\,(n+C)$) during the heating phase, the resulting expression for TL intensity I becomes

$$I = \frac{sA_h}{A_n N}(n+C) n \exp\left(-\frac{E}{kT}\right) \tag{3.34b}$$

Equation (3.34a) and (3.34b) become identical if $s/(N+C)$ in (3.34a) and sA_h/A_nN in (3.34b) are replaced by a single constant s' having units cm^3 s^{-1} viz

$$I = s'(n+C) n \exp\left(-\frac{E}{kT}\right) \tag{3.34c}$$

It may be seen that the expression (3.34c) may be split into two terms, one containing n and the other containing n^2, which means that the resulting expression is a mixture of FO and SO kinetics. It is therefore named as MO kinetics. It may be noted here that two constraints have been imposed on the physical scheme used to derive the MO kinetics expression (1) $A_n = A_h$, and (2) $n \ll N$ together with $(A_n(N-n)) \gg A_h(n+C)$. The solution of (3.34c) provides the equation for the MO glow curve when the linear heating rate of $dT/dt = \beta$ is used [22], viz

$$I = \frac{s' C^2 \alpha \exp\left[\frac{s'C}{\beta} \int\limits_{T_0}^{T} \exp\left(-\frac{E}{kT'}\right) dT'\right] \exp\left(-\frac{E}{kT}\right)}{\left[\exp\left\{\frac{s'C}{\beta} \int\limits_{T_0}^{T} \exp\left(-\frac{E}{kT'}\right) dT'\right\} - \alpha\right]^2} \tag{3.35}$$

where α is the KO parameter in MO kinetics and is defined as $\alpha = n_0/(n_0 + C)$. Visocekas [24] and Chen et al. [22] state that C may be positive or negative. Sunta et al. [22] contest this view and state that a negative value of C leads to a number

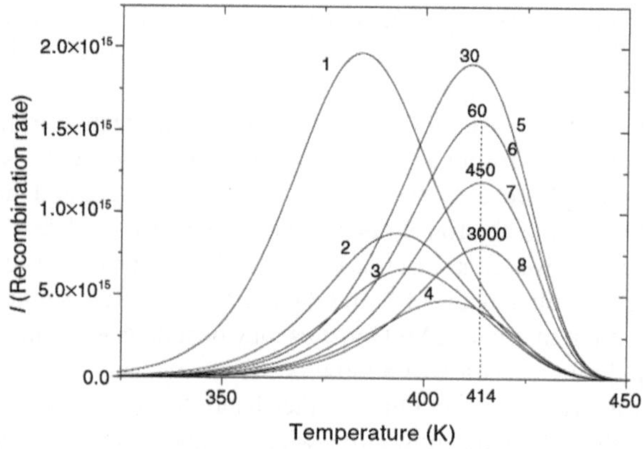

Fig. 3.15 Glow curves of MO kinetics for different initial trap occupancies n_0/C. Input values used are $E = 1$ eV, $s' = 10^{-5}$ cm^3 s^{-1}, $C = 10^{16}$ cm^{-3}, $N = 10^{18}$ cm^{-3} and $\beta = 1$ Ks^{-1}. Glow curves 1–8 are for $n_0/C = 9, 4, 3, 1, 0.25, 0.1, 0.01$ and 0.001 respectively. These correspond to $\alpha = 0.9, 0.8, 0.75, 0.5, 0.2, 0.091, 0.0099$ and 0.000999 respectively. Intensities of the glow curves 5–8 are multiplied by the factors shown respectively over the curves. The value of $s' = 10^{-5}$ cm^3 s^{-1} is based on $N = 10^{18}$ cm^{-3}, $C = 10^{16}$ cm^{-3}, $s = 10^{13}$ s^{-1} and $A_n = A_h$ [22, Fig. 7]

of inconsistencies. For example, a negative C when $|C| > n_0$ would produce negative α and negative TL intensity. Also as a consequence of negative value of C, the population h of RC becomes less than that of the active traps n. This would violate the charge neutrality of the sample. Further, the condition $h < n$ leads to the abrupt ending of the glow peak [16], a situation never seen in the real samples. Shenker and Chen [26] also show that under this condition the E value may turn out to be negative. Thus, a negative C on all accounts appears to be unphysical. Although the proponents of the MO model have used C as an arbitrary constant, Sunta et al. [22] have given a physical meaning to its positive value by using $C = M$, where M is the population of the thermally disconnected deep traps (TDDT). The TDDTs may be assumed to be totally filled, so that they do not capture the carriers liberated from the active traps during heating. Because of the existence of the TDDTs, additional RCs, numbering M would be present to maintain the charge neutrality of the sample. Thus the population of RC would become equal to $(n + M)$ and C may then be replaced by M in the MO model, giving the arbitrary constant C in Visocekas' model, a physical meaning.

Typical glow curves of MO kinetics model are shown in Fig. 3.15. As may be seen they shift to higher temperatures as n_0/C decreases. This behavior is like that of SO and GO kinetics model (compare with Figs. 3.8 and 3.11). However, unlike SO and GO whose glow peaks continue to shift indefinitely with decreasing initial trap occupancy n_0/N, in the case of MO, the glow peaks attain a fixed position at $n_0/C \leq 10^{-1}$. Under this condition, the glow peaks attain FO kinetics. On the other hand at $n_0/C \geq 10$ the glow peaks attain SO kinetics. This behavior may be easily understood from expression (3.34c). This shows the constraint in the MO kinetics model that this model may be obtained only under a narrow range of $0.1 \leq n_0/C \leq 10$. A major drawback of MO model is that its KO parameters α is dependent on n_0, since as defined above, $\alpha = n_0/(n_0 + C)$. It means that no unique value of α may be assigned to a given glow peak of a sample. It means that MO kinetics provides no better alternative to the GO kinetics model for the glow curve fitting, since in both of those models their KO parameter is dependent on the sample dose. Yossian and Horowitz [27] suggested that MO model is preferable to GO, because the former is based on a physical model, whereas the latter is simply empirical. However, as discussed above, MO kinetics has its own draw backs. Notwithstanding the deficiencies of both GO and MO kinetics, Sunta et al. [22] state that a significant point in favor of MO kinetics may be that its KO parameter α remains constant at all T during emission of a glow peak, since n_0 and C are constant for any given sample at a any give dose. On the other hand, the KO parameter b of the GO model changes with T along the glow curve when b value is between 1 and 2. However, both α and b change with the trap occupancy (dose), so that no fixed value of either b or α may be assigned to a given glow peak when these models are applied to the glow curves of a real sample at different doses.

3.5 Realistic Physical Models

3.5.1 *Generalized Scheme*

In the preceding sections, we have dealt with simple models. The first three of these models namely the RW [2], the GG [7] and the MP (GO) [8] models are essentially based on one-trap-one RC (OTOR) scheme. The fourth namely the MO model is a simple modification of the OTOR model in which an extra parameter designated by a constant C is arbitrarily added to the population of RCs. The constant C has later been given physical meaning in terms of the deeper level traps by Sunta et al. [22] as described in the preceding section. Chen and Pagonis [28] have simulated glow curves for a more complex system consisting of five types of traps and five types of RC. The untrapped carriers from each type of the trap have option to undergo recombination with any of the RC. Their results show that the traps which empty earlier (because of lower activation energy) produce glow peaks of FO kinetics and those which empty towards the end of the heating to read the TL may produce glow peaks of GO, some even with a long falling tail which would mean the glow peak having kinetics order greater than 2. This happens because of significant accumulation of free electrons in the conduction band due to the relaxation rates (recombination and retrapping) not being able to cope up with the excitation rate at high temperatures. This though, depends on concentration and the cross section values of the input parameters (Sunta et al. [29]).

In the background of the information that we have gathered from these various models, we may now look at the TL of real materials. In real materials, there are

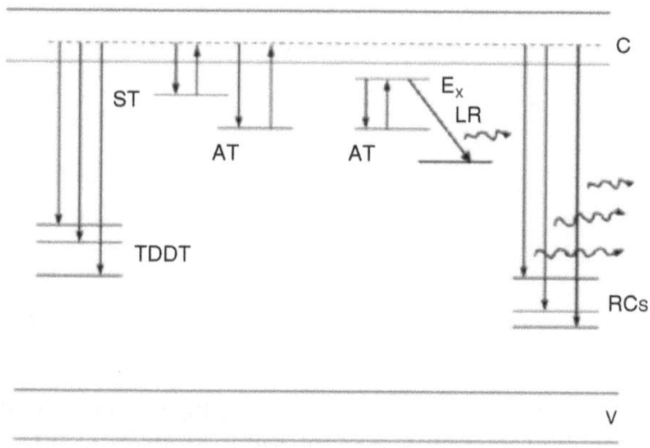

Fig. 3.16 Generalized energy level scheme for TL. AT, ST, and TDDT show the active traps, the shallow traps and the thermally disconnected deep traps respectively, *RC* are the recombination centers, *LR* means local recombination. Thermal excitation of the trapped electron is shown by the *upward vertical arrow* and the relaxation routs of the free electron are shown by *downward ones*

multiple types of traps and multiple types of RCs as evidenced by the appearance of a number of glow peaks in their glow curves and often a number of different wave lengths in the spectra of the TL glow. One may, for example, see the case of fluorite mineral which shows as many as about 15 glow peaks (Sunta [30]) from −125 to 700 °C and a variety of rare earth impurities as RC (Sunta [31]). A general energy level scheme commensurate with these characteristics of the TL materials is given in Fig. 3.16. This figure shows that the electrons ejected out of the traps have multiple options for relaxation. These options may be divided into three types: (1) undergo recombination with any of the RCs to emit TL, (2) be retrapped into the vacant traps of any type shallower than the active traps. These retrapped electrons may be re-excited and recombine with the RCs later to produce the TL during the heating, (3) be recaptured in the thermally disconnected deeper traps (TDDT), from where they are not re-excited during emptying of the given active trap. The options (2) and (3) are nonradiative transitions. The figure shows the various transition routes. We may call this scheme the delocalized recombination model (DRM). The RW, GG, MP and MO models also come under the DRM category. The system of multiple types of traps described above in which thermally disconnected deep traps can recapture the charge carriers untrapped from the active traps has been named as interactive multi-trap system (IMTS) by Sunta et al. [32]. If the deep traps are pre-filled till saturation, these would not recapture the thermally liberated charge carriers, the system then is called the non-interactive multi-trap system (NMTS).

Another plausible model for TL is the local recombination model (LRM). In this model, the trap and the RC are spatially associated (see Fig. 3.17). The charge carrier from the ground level of the trap is excited thermally to a higher bound state from where it may recombine with the spatially associated RC to emit TL or

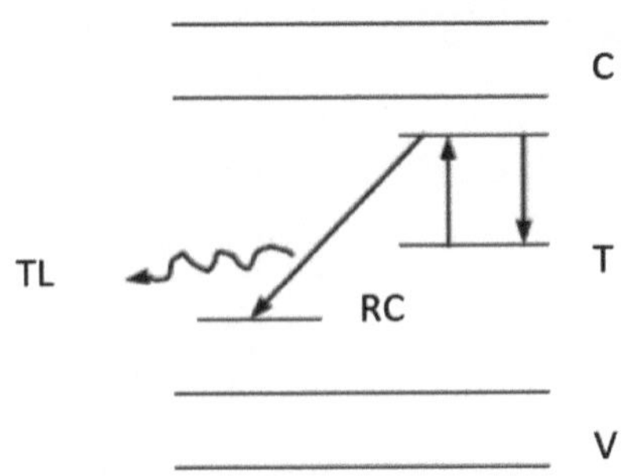

Fig. 3.17 Energy level scheme for localized recombination

it may fall back into its original position (retrapping) to be re-excited later until finally it undergoes the recombination transition (Sunta et al. [33]). A major difference between the DRM and the LRM is that in the former one may see the thermally stimulated electrical conductively (TSC) peaks during TL emission where as in the latter one does not expect the TSC peaks, since in the latter case untrapped electron is not thrown into the conduction band during the heating. There are a number of experimental evidences to support that the traps and the RC may exist in close vicinity to each other or may be associated spatially in some form or other (see Yang et al. [34], Townsend and Rowlands [35]). On the other hand, there exist experimental results in a variety of real materials which generally show that the untrapped charge carriers are delocalized during the heating. This is evidenced by the thermally stimulated conductivity (TSC) peaks appearing simultaneously with the TL peaks in some materials (see Brind et al. [36], Yukihara et al. [37]). Another phenomenon which supports the DRM is the photo transfer of TL (PTIL) (see Sunta et al. [38]). In PTTL, the photonically excited carriers from the traps are re-captured in various other types of traps which are vacant. This may happen only when the untrapped carriers are delocalized before being recaptured by other types of traps. From this short discussion, it is apparent that both the routes of recombination namely delocalized and localized may be taking place in the TL emission process, the predominance of one over other may vary from material to material. Also in the same material some glow peaks may result from the local recombination and some others from the delocalization mechanism. Some authors (Mandowski [39]). Kumar et al. [40] have assumed a scheme in which the charge carriers in the thermally excited state of the trap may have two options namely either to undergo recombination as envisaged under LRM or they may be further raised to the conduction band to get delocalized before recombination. Thus the TL glow curve from a given trap may be a mixture produced by two routes of recombination namely localized as well as delocalized. Figure 3.18 shows the transitions in this scheme.

The schemes discussed above are not simple for developing analytical expressions. It has been a usual practice to deal with these numerically. Also numerical technique provides better insight into the whole process of charge transport during the heating phase. The rate equations are used to describe the transport of the thermally excited carriers to different relaxation routes. Let us

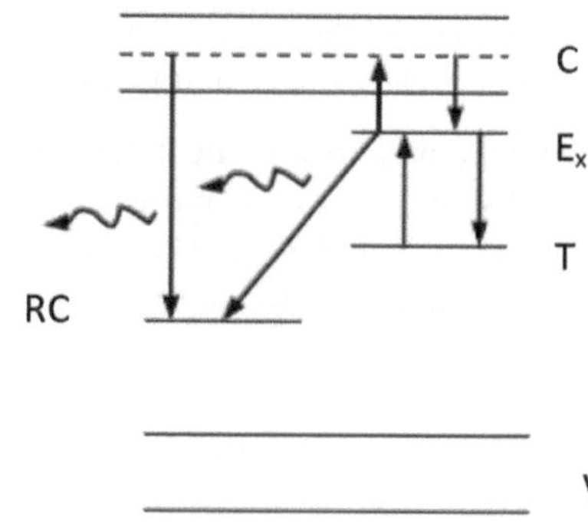

Fig. 3.18 Energy level scheme for TL arising simultaneously from localized and delocalized recombination. TL emission is shown by the *arrow headed wavy line*

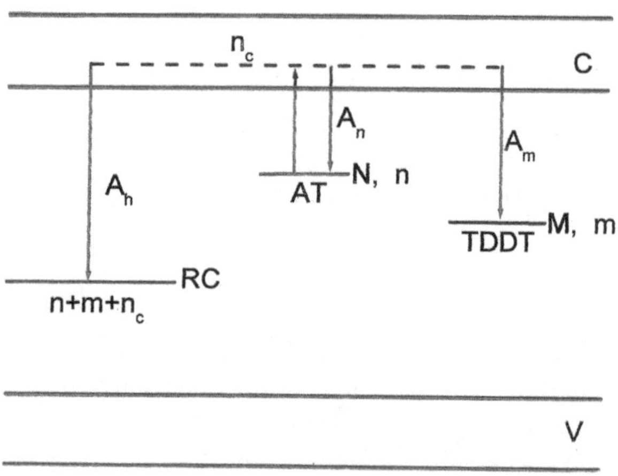

Fig. 3.19 Simplified energy level scheme of the generalized model. It consists of one type of traps, one type of RCs and one type of TDDTs

assume that the individual types of traps are energetically well separated. One may then consider that only one type of traps is thermally active at a time. In the following paragraphs, we consider two types of schemes under the DRM, namely those involving interactive deep traps and those in which the deep traps are noninteractive.

3.5.2 Interactive Multi-trap System

For the sake of simplicity, we consider in the following only one type of active traps (ATs), one type of RCs and one type of TDDTs (see Fig. 3.19). This means we are dealing with a glow curve with single peak. The various terms in the rate equations for this scheme are

excitation rate, $R_{ex} = ns\,\exp\left(-\frac{E}{kT}\right)$
retrapping rate, $R_{ret} = A_n\,(N - n)n_c$

recombination rate, $R_{\mathrm{rec}} = A_h\,(n + m + n_c)\,n_c$
recapture rate in deeper traps, $R_{\mathrm{recap}} = A_m\,(M - m)\,n_c$

where M and m are the total and the filled concentrations respectively of the TDDTs. A_m is the capture coefficient of the TDDT. The meaning of other parameters is already explained in the earlier sections. (refer also Appendix-A and B.) The equations for the n, m, and n_c as a function of time, then are

$$\frac{\mathrm{d}n}{\mathrm{d}t} = -ns\,\exp\left(-\frac{E}{kT}\right) + A_n(N - n)\,n_c \tag{3.36}$$

$$\frac{\mathrm{d}m}{\mathrm{d}t} = A_m(M - m)n_c \tag{3.37}$$

$$\frac{\mathrm{d}n_c}{\mathrm{d}t} = ns\,\exp\left(-\frac{E}{kT}\right) - A_n(N - n)n_c - A_m(M - m)n_c - A_h(n + m + n_c)n_c \tag{3.38}$$

Intensity of TL at any instant is given by the recombination rate at that instant, viz

$$I = R_{\mathrm{rec}} = A_h(n + m + n_c)\,n_c \tag{3.39}$$

The concentration of RC is equal to $(n + m + n_c)$ so that it balances with the sum of the concentrations of the filled active traps n, filled TDDT m and the free electrons n_c in the conduction band.

Sunta et al. [32] have computed the TL glow curves for selected sets of parameters using the above rate equations. Calculations were carried out numerically. Figures 3.20, 3.21, 3.22 show the glow curves for three different parametric combinations to examine the changes in peak temperature T_m and the peak shape factor δ/ω due to the changes in the parameter values. The values of E, s, N, and A_n are kept same in all of them. These are $E = 1$ eV, $s = 10^{12}\ \mathrm{s}^{-1}$, $N = 10^{12}\ \mathrm{cm}^{-3}$ and $A_n = 10^{-7}\ \mathrm{cm}^3\ \mathrm{s}^{-1}$. The heating rate β also is kept same, namely $\beta = 1\ \mathrm{K\ s}^{-1}$. The values of M, A_m, and A_h are changed, to see the effect of these changes. Further, the calculations are repeated by changing the initial concentrations n_0 and m_0 of the active traps and the TDDT respectively. The purpose of this is to see if the glow peaks shift their position with trap occupancy like in SO, GO, and MO kinetics and also to examine the effect on the shape factor of the glow peaks which gives the KO value. In other words this exercise allows us to examine the glow peaks of the IMTS model in the perspective of the conventional models of FO, SO, GO, and MO kinetics. The initial concentrations n_0 and m_0 of the ATs and the TDDTs are arrived at by assuming that the traps fill with dose according to the saturating exponential function (see Sunta et al. [41]) and refer to (6.4) and (6.5) in Chap. 6) in which the filling rate constants for ATs and the TDDTs are assumed to be proportional to A_n and A_m, respectively. Figure 3.20 shows that when the traps

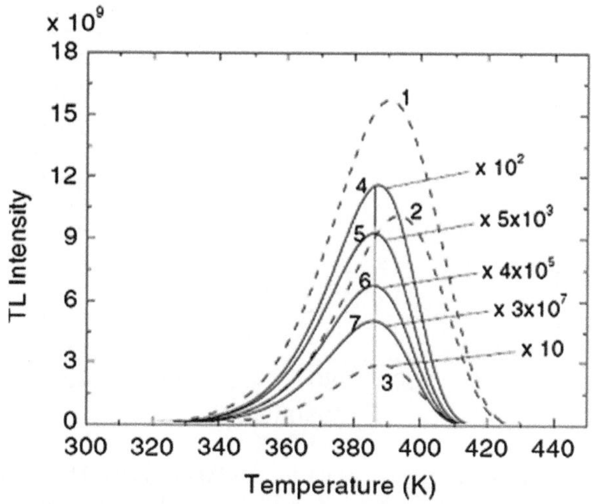

	Peak No						
	1	2	3	4	5	6	7
n_o/N	0.632	0.394	0.095	0.067	0.01	0.001	0.0001
m_o/M	1	0.993	0.632	0.5	0.1	0.01	0.001
$T_m(K)$	390	393	388	387	386	386	386
δ/ω	0.47	0.45	0.43	0.42	0.41	0.41	0.41

Fig. 3.20 Numerically calculated TL glow peaks for the IMTS model for different trap occupancies. The parameter values used in the rate equations are $E = 1$ Ev, $s = 10^{12}$ s^{-1} $A_n = 10^{-7}$ cm^3 s^{-1}, $A_m = 10^{-6}$cm^3 s^{-1}, $A_h = 5 \times 10^{-8}$ cm^3 s^{-1}, $N = M = 10^{12}$ cm^{-3}, $\beta = 1$ K s^{-1}. The initial values of the trap occupancies n_0 and m_0 are obtained by assuming that the traps fill up with dose according to a saturating exponential function (refer (6.4) and (6.5) in Chap. 6) and their filling rate constants are proportional to A_n and A_m, respectively. The input values of n_0/N, and m_0/M, the resulting values of temperature of the peak maxima T_m and the symmetry factor δ/ω are tabulated above for each computed glow curve. The intensities at lower trap frequencies are multiplied for clarity and the respective multiplication factors are indicated in the figure. It is seen that when $m_0/M \leq 0.5$ and $n_0/N \leq 0.067$ the glow peaks obtained are characteristically of first order. These are shown with full curves [32, Fig. 2] © IOP Publishing Ltd, reproduced by permission, all rights reserved

AT and the TDDT are filled up to no more than 10 % of their saturation (n_0/N and $m_0/M \leq 1/10$), the glow peaks demonstrate a fixed T_m and a fixed shape like that of FO kinetics. At greater than 10 % occupancy in either AT or TDDT or in both, the peaks shift their position as well as change their shape. The values of T_m and shape factor δ/ω are listed in the figure captions. In Fig. 3.21 in which M is greater than N by a factor of 10^2, and A_n, A_m, and A_h are all equal (=10^{-7} cm^3 s^{-1}), the glow

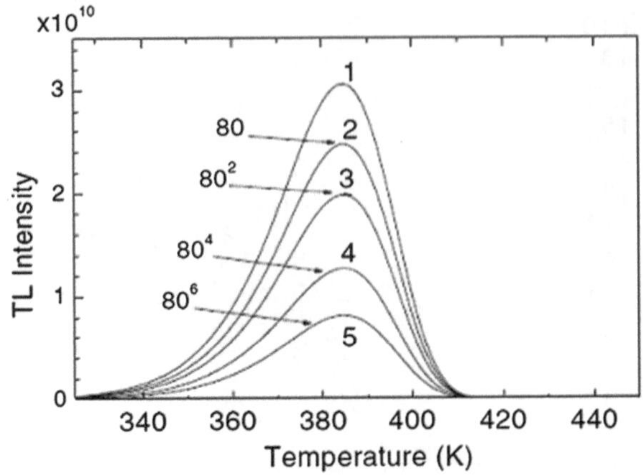

Fig. 3.21 Numerically-computed TL glow peaks for the IMTS model. The parameters E, s, N, A_n and β have the same values as in Fig. 3.20. $M = 10^{14}$ cm^{-3}, $A_m = A_h = 10^{-7}$ cm^3 s^{-1}. Glow peaks 1–5, respectively, are for $n_0/N = m_0/M = 1$, 10^{-1}, 10^{-2}, 10^{-4} and 10^{-6}. $T_m = 385$ K and $\delta/\omega = 0.42$ for all the five peaks. The intensity multiplication factors used for plotting the curves are shown for each glow peak [32, Fig. 3]

peaks appear at the same T_m and have same shape for all trap occupancies like in the case of FO kinetics. In Fig. 3.22 in which A_h is increased by a factor of 20, in comparison to that used in Fig. 3.20, with all other parameters being same as in Fig. 3.20, the glow peaks again have a stable temperature (T_m) and stable shape at all initial trap occupancies n_0/N and m_0/M. The shape factor for all these glow peaks is about 0.42. The reason for the effect of changes in parameter values described here may be better understood from the analytical expression that we may derive for the IMTS model used in this exercise.

Using (3.36), (3.37), and (3.39) in (3.38) we may write (3.38) as

$$\mathrm{I} = -\frac{dn}{dt} - \frac{dm}{dt} - \frac{dn_c}{dt} \tag{3.40}$$

Now let us apply the QE conditions, namely

$$n_c \ll (n + m) \tag{3.41}$$

and

$$\left|\frac{dn_c}{dt}\right| \ll \left|\frac{dn}{dt} + \frac{dm}{dt}\right| \tag{3.42}$$

Equation (3.38) may then be written as:

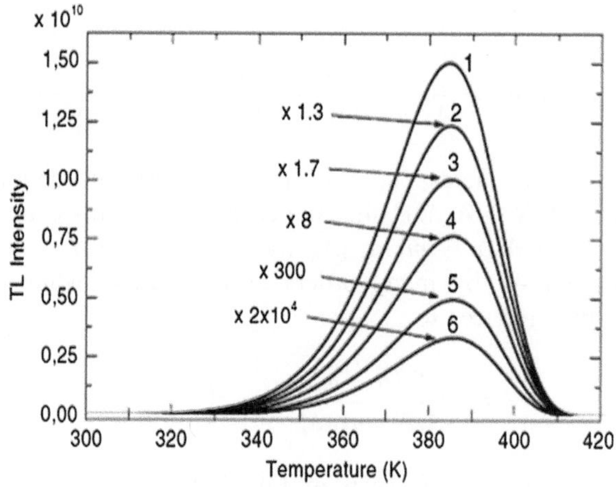

Peak No						
	1	2	3	4	5	6
n_o/N	1	0.632	0.394	0.095	0.01	0.001
m_o/M	1	1	0.993	0.632	0.1	0.01
T_m(K)	385	385	385	386	386	386
/	0.42	0.42	0.42	0.42	0.42	0.42

Fig. 3.22 Numerically-computed TL glow peaks for the IMTS model at different trap occupancies values of all the parameters are the same as in Fig. 3.20 except that $A_h = 10^{-6}$ cm^3 s^{-1}. Values of n_0/N, m_0/M, T_m and δ/ω are listed above for each peak. Multiplication factors used for plotting the TL intensities are shown in the figure for each glow peak [32, Fig. 4] © IOP Publishing Ltd, reproduced by permission, all rights reserved

$$A_h(n + m)n_c = ns\exp\left(-\frac{E}{kT}\right) - A_n(N - n)n_c - A_m(M - m)n_c$$

This yields the expression for n_c viz

$$n_c = \frac{ns\exp\left(-\frac{E}{kT}\right)}{A_h(n + m) + A_n(N - n) + A_m(M - m)} \tag{3.43}$$

and for TL intensity I viz

$$I = A_h(n+m)n_c = \frac{A_h(n+m)\,ns\exp\left(-\frac{E}{kT}\right)}{A_h(n+m)+A_n(N-n)+A_m(M-m)} \tag{3.44}$$

At low initial occupancies ($n_0 \ll N$ and $m_0 \ll M$) the denominator in this equation becomes ($A_nN + A_mM$), since under these low trap occupancy conditions we would also have $A_h(n + m) \ll [A_n\ (N - n) + A_m\ (M - m)]$ unless A_h is greater than A_n and A_m by many orders of magnitude, which is least likely. Equation (3.44) thus reduces to

$$I = \frac{A_h(n+m)}{A_nN + A_mM} ns\exp\left(-\frac{E}{kT}\right) \tag{3.45}$$

At low-trap occupancies as seen in Fig. 3.20 most of the charge carriers released from the active traps during heating are recaptured in the TDDTs. For example in the case of glow curve 4 the recombination rate is about two orders of magnitude lesser than the recapture rate and it reduces progressively as the trap occupancy further goes down in glow curves 5–7. The retrapping rate is one order of magnitude lesser than the recapture rate. This is because we have $A_m \gg A_n, A_h$ in this case. The result is that at low trap occupancies any decrease in n during heating is compensated by an almost equal increase in m with the result that $(n + m) \approx (n_0 + m_0)$ and (3.45) then reduces to

$$I = \frac{A_h(n_0+m_0)}{A_nN + A_mM} ns\exp\left(-\frac{E}{kT}\right) \tag{3.46}$$

This is a FO TL equation representing glow curves 4–7 in Fig. 3.20. As stated before these glow curves were calculated numerically using the rate equations (3.36)–(3.39). In IMTS model under certain conditions it is possible to get FO kinetics at all trap occupancies including high and low. The essential condition is (a) $A_m\,M \gg A_n\,N$. The additional conditions are either (b) $M \gg N$ as used for the parameters in Fig. 3.21 or (c) A_h and $A_m \gg A_n$ together with $M \geq N$ as used for the parameter set of Fig. 3.22. When the conditions (a) and (b) hold, one gets $m_0 \gg n_0$ and $A_m\ m_0 \gg A_n n_0$ at initial trap occupancies which means $m \gg n$ and $A_m\ m \gg A_n\ n$ at all temperatures during heating. Further, the value of m remains practically constant at m_0, since with $m_0 \gg n_0$ there would be no significant increase in m_0 during heating due to recapture of free carriers untrapped from AT. Equation (3.44) thus reduces to the FO form, viz

$$I = \frac{A_h m_0 ns\exp\left(-\frac{E}{kT}\right)}{A_mM + m_0(A_h - A_m)} \tag{3.47}$$

This is a FO TL equation applicable to all trap occupancies.

When the conditions (a) and (c) apply, one gets FO kinetics at high trap occupancies because of $R_{rec} \gg R_{ret}$, R_{recap} and at low trap occupancies it is because of $R_{recap} \gg R_{ret,}$ In the former case (3.44) reduces to

$$I = ns\exp\left(-\frac{E}{kT}\right) \tag{3.48}$$

and in the latter case one gets

$$I = \frac{A_h m_0}{A_m M} ns\exp\left(-\frac{E}{kT}\right) \tag{3.48a}$$

In this equation only m_0 appears for the number of RC, because not enough of AT get filled up during irradiation due to $A_n \ll A_m$, and so that we get $n_0 \ll m_0$.

The above discussion shows that the common condition for arriving at the analytical equations of FO kinetics is that the recombining fraction F should be a constant. The three equations from (3.46) to (3.48a) belong to FO kinetics but they differ in respect of their recombining fraction F. Except for (3.48) in which case $F = 1$ like in RW model, the value of F in the other equations is much smaller than 1 and depends on the trap occupancies. The effect of this leads to super-linear growth of TL intensity. The super-linearity arises because of non-radiative capture of untrapped carriers from AT by the vacant positions in TDDTs at low trap occupancies. This is dealt in greater detail in Chap. 6.

To sum up it is seen that, (1) under certain parametric conditions, the TL of IMTS model is of FO kinetics only at low initial trap occupancies. (2) Under certain other parametric combinations as described above FO kinetics is achieved both at high and low trap occupancies. An important common condition is that the total concentration of TDDT should be much greater than that of the active traps ($M \gg N$). The condition $M \gg N$, in fact, may naturally be present in real materials, since a number of TDDTs at various deeper levels are always expected in real materials (refer to Sect. 3.7 in this chapter). Another important condition as stated above is $A_m M \gg A_n N$, the effect which is to produce $m_0 \gg n_0$. The effect of these conditions is seen in (3.47), (3.48), and (3.48a).

3.5.3 Non-interactive Multi-trap System

A number of earlier authors (see Lewadowski and McKeever [3], Dussel and Bube [16], Saunders [17], Shenker and Chen [26], Kelly et al. [42], Mckeever et al. [43] and Lewadowski et al. [44]) have arbitrarily assumed that the TDDTs do not recapture the free charge carriers during the TL read out heating. We call such a scheme as non-interactive multi-trap system (NMTS). In the framework of the general scheme described above under IMTS, this can happen if the TDDTs are pre-filled until saturation ($m_0 = M$) or if $A_m = 0$ in (3.44). Assuming that this

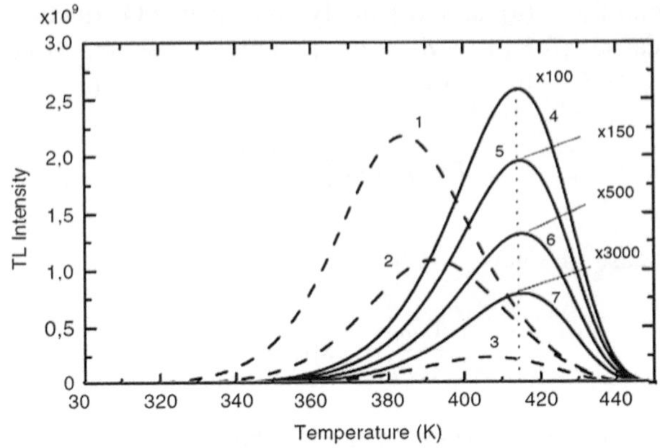

	Peak No.						
	1	2	3	4	5	6	7
n_0/N	1	$5*10^{-1}$	10^{-1}	10^{-2}	$5*10^{-3}$	10^{-3}	10^{-4}
T_m	384	392	404	414	415	415	415
δ/ω	0.53	0.52	0.46	0.42	0.42	0.42	0.42

Fig. 3.23 Numerically computed Glow curves of NMTS model at different n_0/N. Input parameter values E, s, A_n, A_m, A_h and β are same as in Fig. 3.21. $N = 10^{11}\ \text{cm}^{-3}$ and $M = 10^{10}\ \text{cm}^{-3}$. Intensity values of some of the glow curves are multiplied for plotting by the factors as shown. Values of n_0/N, T_m and δ/ω for each of the glow curves are listed above [32, Fig. 5]

constant population of TDDTs is equal to M, the population of RCs becomes equal to $(n + M)$. For NMTS, the general (3.44) then reduces to

$$I = \frac{A_h(n+M)ns\exp\left(-\frac{E}{kT}\right)}{A_h(n+M)+A_n(N-n)} \tag{3.49}$$

This is the analytical equation for the general case of the NMTS model. Glow curve characteristics however, change with the change in the parameter values like in the case of IMTS model. Glow curves for a case in which $MA_h \ll NA_n$ are given in Fig. 3.23. The input parameter values are given in the figure caption. Equation (3.49) reduces to FO at low trap occupancies ($n \ll N$ and M) viz

$$I = \frac{A_hM}{A_hM + A_nN} ns\exp\left(-\frac{E}{kT}\right) \tag{3.50}$$

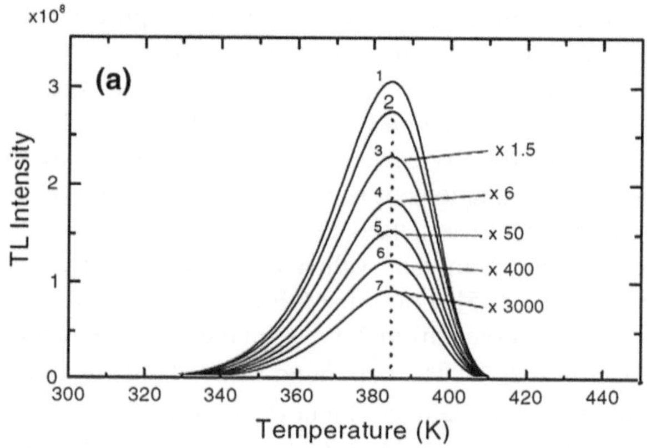

Fig. 3.24 Numerically computed glow curves of NMTS model for the condition $A_hM \gg A_nN$. The values of E, s, A_n and β are the same as in Fig. 3.21. $N = 10^{10}$ cm^{-3} and $M = 10^{11}$ cm^{-3} and $A_h = 10^{-6}$ cm^3 s^{-1}. Glow curves are shown for $n_0/N = 1$, 0.9, 0.5, 10^{-1}, 10^{-2}, 10^{-3}, and 10^{-4}. They all have the same T_m (=385 K) and same δ/ω (=0.41) implying that all the seven peaks are of first order kinetics. Intensities of glow peaks 3–7 are multiplied by the factors as shown for plotting [32, Fig. 6]

Glow peaks marked 4–7 in Fig. 3.23 show that FO is obeyed when $n_0 < N/10$. If we have $A_hM \gg A_nN$, (3.50) reduces exactly to the RW form viz

$$I = ns\exp\left(-\frac{E}{kT}\right)$$

Glow curves for a given set of parameters under the condition $A_hM \gg A_nN$ are shown in Fig. 3.24. In this case it is not necessary to have low trap occupancy to get FO kinetics. Like in the case of Figs. 3.21 and 3.22, the glow curves are of FO type at all trap occupancies. As stated before all these glow curves are calculated numerically using the respective rate equations of the NMTS and IMTS models.

Two important points emerge from this discussion; (1) the glow peaks of both NMTS and IMTS models attain FO kinetics (Subject to QE conditions being satisfied), when the trap occupancies are low ($n_0/N < 0.1$ in NMTS and n_0/N, $m_0/M < 0.1$ in IMTS). (2) FO kinetics is obtained in both of these models at all trap occupancies if the TDDTs are in greater abundance than the ATs ($M \gg N$ and $A_mM \gg A_nN$ in IMTS and $A_hM \gg A_nN$ in *NMTS*). From these results it may be seen that except under the limited conditions $N \geq M$ and $A_nN \geq A_mM$, under which FO is obtained only when the trap occupancy is low, (n_0/N, $m_0/M < 0.1$), one gets FO kinetics in these models at all trap occupancies. However the condition $N \geq M$ is not likely to occur in real materials, since as stated before one expects a variety of defect centers at deeper levels, which may account for higher values of M in comparison to N.

In the above discussion, conclusions regarding the prevalence of FO kinetics are drawn using the analytical equations which are derived by assuming the QE condition in the rate equations. The glow curves used in support of the conclusions however, are computed using the rate equations of these models without the use of QE conditions. For the validity of using these glow curves to support the conclusions from analytical equations, it is essential that the QE condition be satisfied under the parametric conditions used in these numerical calculations. The QE condition depends on the numerical values of the parameters. A simple criterion developed by Sunta et al. [45] shows that QE is well satisfied when the life time τ of the free carriers in the delocalized band is smaller than the recording time of the glow peak by a factor $\leq 10^{-3}$. The maximum value of τ is $\frac{1}{MAh}$ in the NMTS model and $1/M\,A_m$ in the IMTS model (see Sunta et al. [45]). The glow curve recording time in Figs. 3.20, 3.21, 3.22, 3.23, 3.24 is about 100 s. This means that the τ values which are less than 10^{-1} s would satisfy the QE criterion. In these cases the values of M, A_h, and A_m are such that this condition is easily satisfied.

We have dealt with some analytical models as well as computed the glow curves of simple models namely the interactive and noninteractive system of traps using the rate equations of the models and numerically calculated the glow peaks without assuming the QE approximation. Results of these computations show under what parametric conditions FO kinetics prevails in the TL. This helps us in finding the answer to the question why FO kinetics is seen to invariably prevail in the TL of real materials. In the foregoing we have dealt exclusively with the theory and behavior of the TL glow peaks under different parametric conditions. For a wider view on the range of associated phenomena the reader may refer to Chen and Pagonis [46]. These authors have dealt with the various sister phenomena like thermally stimulated conductivity (TSC), optically stimulated luminescence (OSL), thermally stimulated electron emission (TSEE) and radiophotoluminescence (RPL) together with their theory, applications and simulations in some procedures.

3.6 Localized Recombination Model

As mentioned before, when the trap and the RC exist in the same defect complex or in close vicinity to each other, recombination of the thermally excited charge carrier may occur without getting delocalized. In Fig. 3.17 is shown the energy level scheme for this model. The charge carrier transport during heating to read the TL may be described by the following equations.

$$\frac{dn}{dt} = -ns\exp\left(-\frac{E}{kT}\right) + sn_e \tag{3.51}$$

$$\frac{\mathrm{d}n_e}{\mathrm{d}t} = ns\exp\left(-\frac{E}{kT}\right) - sn_e - vn_e \tag{3.52}$$

where n_e is the concentration of the charge carriers in the excited state of the trap E_x. Following Halperin and Braner [15] the retrapping rate constant is assumed equal to the excitation rate constant s. This, according to these authors, follows from the principle of detailed balance. The parameter s in the present case does not represent the frequency factor s defined in Sect. 3.2.1, since the electron is not delocalized into the conduction band. The TL intensity I is given by vn_e, where v is the recombination rate constant. From (3.52), we get

$$n_e = \frac{1}{s+v}\left[ns\exp\left(-\frac{E}{kT}\right) - \frac{\mathrm{d}n_e}{\mathrm{d}t}\right] \tag{3.53}$$

Assuming $\frac{\mathrm{d}n_e}{\mathrm{d}t} \ll ns\exp\left(-\frac{E}{kT}\right)$ in this equation, which in any case is expected to be true, as seen from (3.52), we get

$$I = vn_e = \left(\frac{vs}{s+v}\right)n\exp\left(-\frac{E}{kT}\right) \tag{3.54}$$

If we use $\frac{vs}{s+v} = s'$ in (3.54), it becomes similar to the RW equation with s replaced by s'. It is thus seen that in LRM, the TL is always of FO kinetics irrespective of whether recombination rate is dominant ($v \gg s$) or retrapping rate is dominant ($s \gg v$). The equation for the glow curve accordingly is

$$I(T) = n_0 s'\exp\left(-\frac{E}{kT}\right)\exp\left[-\int_{T_0}^{T}\frac{s'}{\beta}\exp\left(-\frac{E}{kT'}\right)\mathrm{d}T'\right] \tag{3.55}$$

To arrive at (3.54), earlier workers (see Chen and McKeever [13], Chen [47], Bull [48]) have used the same assumptions for QE as applicable to DRM viz $n_e \ll n$ and $\left|\frac{\mathrm{d}n_e}{\mathrm{d}t}\right| \ll \left|\frac{\mathrm{d}n}{\mathrm{d}t}\right|$ or $\frac{\mathrm{d}n_e}{\mathrm{d}t} \approx 0$. Sunta et al. [33] state that these are not relevant in LRM. This is because: (i) unlike in the models which come under DRM, in the present model the total population of RC ($h = n + n_e$) does not come into picture for calculating the recombination rate, since the charge carrier in the excited state E_x of a trap recombines only with that RC which is spatially associated with the trap. Thus $n_e \ll n$ is not relevant in LRM. (ii) It may be seen from (3.51) and (3.52) that under the retrapping dominant conditions ($s \gg v$), we get $\left|\frac{\mathrm{d}n_e}{\mathrm{d}t}\right| \approx \left|\frac{\mathrm{d}n}{\mathrm{d}t}\right|$, yet (3.53) remains valid and the TL equation retains the FO form with the PF $s' = v$. It is only under the recombination dominant condition ($v \gg s$) that we would have $\left|\frac{\mathrm{d}n_e}{\mathrm{d}t}\right| \ll \left|\frac{\mathrm{d}n}{\mathrm{d}t}\right|$ and the s' value in (3.54) in that case becomes equal to the excitation rate constant s. Sunta et al. [33] have numerically calculated the ratio

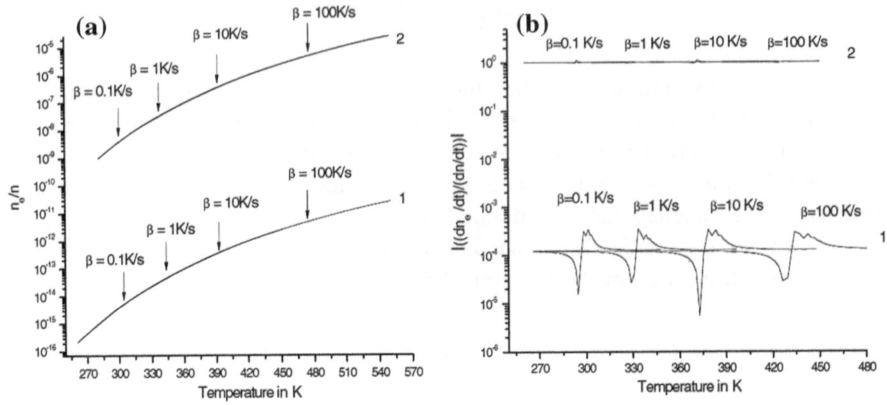

Fig. 3.25 **a** The ratio n_e/n versus T for $s = 10^6$ and $\nu = 10^{12}\ \mathrm{s}^{-1}$ shown as curve 1. For $s = 10^{12}$ and $\nu = 10^6\ \mathrm{s}^{-1}$, it is shown as curve 2. *Vertical arrows* point towards the glow peak temperature T_m value for the four glow curves (not shown here) recorded by using the heating rates indicated over the arrows. **b** $|dn_e/dt|/|dn/dt|$ versus T for the stated pairs of s and ν shown as curves 1 and 2 respectively. The glow curves for both pairs of s and ν are coincident for all the four heating rates used here. The other input parameters for the glow curves used here are $E = 0.5$ eV and $n_0 = 10^{14}\ \mathrm{cm}^{-3}$ [33, (a) Fig. 4 and (b) Fig. 5] © IOP Publishing Ltd, reproduced by permission, all rights reserved

$\left|\frac{dn_e}{dt}\Big/\frac{dn}{dt}\right|$ for the two cases and find that this ratio is equal to 1 in retrapping dominant (s $\gg \nu$) cases and is far smaller than 1 (the value depending on s/ν ratio) in recombination dominant cases (see Fig. 3.25). It is simple to understand why $\frac{dn_e}{dt} \approx \frac{dn}{dt}$ in s $\gg \nu$ cases. The reason may be understood by looking at the (3.51) and (3.52). In these cases, both the excitation rate and the retrapping rate are high because of the high value of s. The recombination rate on the other hand is much smaller than both of these. Under this condition the charge carriers of the active traps recycle many times ($\approx \frac{s}{\nu}$ times) between the ground state of the trap and its excited state E_x during heating before finally relaxing into the RC to emit TL. This leads to $\left|\frac{dn_e}{dt}\right| \approx \left|\frac{dn}{dt}\right|$. Also the $^{n_e}/_{n}$ ratio in $s \gg \nu$ cases is much greater than that in $\nu \gg s$ cases. Sunta et al. [33] have calculated the actual values of these ratios as well as the ratios $\left|\frac{dn_e}{dt}\Big/\frac{dn}{dt}\right|$ for a given set of input parameters, both for s $\ll \nu$ and s $\gg \nu$ cases. When s and ν are exchanged in the numerical computation, the glow peaks produced are same for given E and b, since the PF s' value given by $s\nu/(s + \nu)$ remains the same (see (3.54)). The ratios $\frac{n_e}{n}$ and $\left|\frac{dn_e}{dt}\Big/\frac{dn}{dt}\right|$ however, are seen to change by large factors (see Fig. 3.25). The characteristics of the glow peaks of LRM are essentially the same as those of the conventional FO kinetics discussed in Sect. 3.2.1.

3.7 KO of the Glow Curves of Real Materials

Various TL dosimetry phosphors based on impurity doped inorganic insulating or semiconducting materials such as LiF, $Li_2B_4O_7$, Al_2O_3, CaF_2 $CaSO_4$, Mg_2SiO_4, MgO, SiO_2, etc, all show FO behavior, as testified by the invariability of the T_m of their glow peaks with change in dose. A dozen varieties of silicate minerals tested for their dose response too did not show a change in T_m with change in dose [49]. Thus experience with real materials shows that TL in general is a FO kinetics phenomenon. However, a look at published literature shows that often the authors have categorized the TL glow curves of their experimental samples under SO or GO kinetics. This, they do merely on the basis of the peak shape factor μ (see Azorin and Gutierrez [50], Kirsh et al. [51], Dorendra Singh and Ingatomi [52], Tatumi et al. [53], Prokein and Wagner [54], Kirsh et al. [55]). From the applications point of view, if the TL emission were of non-FO kinetics its utility in routine dosimetry and dating would be largely complicated. The TL glow curves of real materials commonly have multiple glow peaks and in some dosimetry phosphors, they grow at different rates with dose. With differing changes in peak positions and peak shapes with change in dose, which would happen if the peaks were of non-FO kinetics, dosimetry would become complicated. The linear relation between the dose and the peak height commonly employed in dosimetry applications would no more hold because of changes in peak width with dose (see Fig. 3.11). The question may then be asked, why is it that FO kinetics is prevalent in the TL emission of various insulating and semi-conducting inorganic materials. In fact as far as our experience goes, we have not come across any material which shows a non-FO behavior by peak shift with change in dose. In RW model [4] FO kinetics is obtained by ignoring retrapping in comparison to the recombination. Retrapping per se, however, cannot be ignored, because if a defect center (trap) captures the free carrier during irradiation, there seems no reason for it to do the same during heating. The only change in condition during heating is the sample temperature, but the latter is not expected to change the capture cross section greatly. The answer, on the basis of the theoretical results, simulations and discussions in the preceding paragraphs, may be given as following: (1) if the emission takes place by local recombination of the thermally excited charge carrier (electron), it is always of FO kinetics. However, as discussed in the beginning of Sect. 3.5, one cannot expect all TL emissions to be by local recombination. (2) most of the authors explain TL emission in terms of delocalized recombination model. In this model, the essential feature, as discussed in Sects. 3.5.2 and 3.5.3, is that the recombining fraction F in the analytical expression of TL should be constant at all temperatures during heating to read the TL signal. Also for the validity of analytical expression QE condition should be satisfied. Both of these conditions, as discussed in Sect. 3.5, depend on the parameter values namely the concentrations (N, M, n_0, m_0) and the capture coefficients (A_n, A_h, A_m) of the active traps and the deep traps. It is shown that the condition for the constant value of F turns out to be; (1) $M \gg N$ and (2) $MA_h \gg NA_n$ in NMTS and $MA_m \gg NA_n$ in

IMTS. The latter conditions involve the coefficients A_n, A_m and A_h. This condition may be violated if $A_n \gg A_m, A_h$. However, A_n, A_m, and A_h are not expected to differ by very large factors whereas the values of M may always be much greater than that of N, since there may exist variety of deep traps at various depths. It therefore means that condition No. 1 may naturally suffice for making the recombining fraction F constant which is the essential condition for obtaining the FO kinetics. The expected range of parameter values M, N, A_n, A_m, and A_h in inorganic TL phosphors is discussed later in Sect. 5.4 which deals with the QE problem. As far as the existence of deeper level traps is concerned, the TL technique is not capable of measuring them or even detecting them because of high background from the heat of the heater pan and also because of thermal quenching of luminescence at high temperatures. Nevertheless, a variety of thermally stable defect centers (traps) may always be envisaged until the sample material approaches its melting point. In terms of their trap depth, these would be at much deeper levels than the active traps for which TL appears at temperatures below about 300 °C, the range usually employed by the TL workers. Notwithstanding the difficulties of high temperature TL measurements, weak glow peaks have been recorded at high temperatures in some of the highly sensitive TL phosphors. For example in blue-green fluorite mineral samples glow peaks have been recorded between 400 and 700 °C (see Al-Kolaly et al. [56]. When these TDDTs are empty the system functions as IMTS with respect to the glow peaks appearing below about 400 °C. When these TDDTs are filled up by pre-irradiation by a high dose (from natural irradiation in the case of a fresh mineral or by subsequent high dose irradiation in the laboratory), the system behaves like the NMTS (see Sect. 3.5). In LiF:Mg, Ti (TLD-100) the traps of the glow peaks appearing at about 400 °C function as TDDT, with respect to the main dosimetry glow peak appearing at about 200 °C. In Al_2O_3:C, electron traps have been recorded which get thermally excited between 800 and 900 °C (see Akselrod and Gorelova [57]).

The question why FO kinetics is prevalent in TL of real materials has been discussed earlier also by McKeever and co-workers (see Lewandowski and McKeever [3], McKeever et al. [43], Lewandowski et al. [44]). They introduced two new functions called $Q(T)$ and $p(T)$ which they define respectively as

$$Q(T) = \frac{R_{\mathrm{ex}} - R_{\mathrm{ret}}}{R_{\mathrm{rec}}} \tag{3.56}$$

and

$$P(T) = \frac{R_{\mathrm{ret}}}{R_{\mathrm{rec}}} \tag{3.57}$$

The function $Q(T)$ is intended to provide an alternative definition to the QE approximations which we have given earlier by (3.41) and (3.42). The function $P(T)$ on the other hand is intended to meaningfully replace the empirical KO parameter b. The above equations may be combined to give

$$\frac{Q(T)}{P(T)} + 1 = \frac{R_{ex}}{R_{ret}} \tag{3.58}$$

Lewandowski and McKeever [3] argue that if one uses $P(T) \gg 1$ together with $Q(T) \approx 1$ this equation leads to $R_{ex} = R_{ret}$, the implication of which is that the charge carriers would just shuttle between the traps and the delocalized band and therefore the traps would not empty and hence no glow peak would be produced. They therefore, conclude that $Q(T) \approx 1$ and $P(T) \gg 1$ do not form a self consistent set of conditions to produce a glow peak. According to this logic a glow curve can be produced only under $P(T) \ll 1$ condition together with $Q(T) \approx 1$. The condition $P(T) \ll 1$, like in RW model, always produces FO kinetics. Thus McKeever and co-workers [3, 43, 44] were led to conclude that the FO case best satisfies the requirements of a realistic system and this in turn explains the apparent dominance of FO kinetics in nature. They further support their argument about the inconsistency of $P(T) \gg 1$ and $Q(T) = 1$ by presenting a $P(T) \gg 1$ case using the parameter values which were too low (Total concentrations of traps and RC each equal to 10^3 cm^{-3}). Such low concentrations particularly of RC fails to provide sufficient vacancies for the relaxation of the untrapped free carriers from conduction band with the result that the carriers tend to get crowded in the conduction band. That is why the QE condition fails and the $Q(T)$ value turns out to be much greater than 1. The available RC being insufficient, the free carriers relax slowly after recycling through re-excitation-retrapping steps. Because of this, the glow emission falls very slowly and TL emission does not have the shape of a glow peak. It may be stressed here that conclusions of general nature cannot be derived from a limited set in particular if the parameter values are unrealistic. Sunta et al. [58] have later shown that the shape of the simulated glow curves as well as those of the $P(T)$ and the $Q(T)$ are totally dependent on the values of the input parameters N, M, n_0, m_0, A_n, A_m, and A_h, irrespective of whether the chosen set of parameters is recombination dominant or retrapping dominant and that the QE approximation ($Q(T) = 1$) may be obtained even under the retrapping dominant condition. Examples of the glow curves of FO kinetics under different parametric combinations with the QE approximation satisfied are already given in Sect. 3.5 above. These include both recombination dominant (see Figs. 3.21, 3.22, and 3.24) and retrapping dominant (see Fig. 3.20 glow peak Nos. 4–7 and Fig. 3.23 peak Nos. 4–7). Main reason why FO kinetics prevails in real materials appears to be the existence of deep traps at temperatures above those of the active traps. These provide for the additional RC. Conclusions of general nature cannot be derived from a limited set in particular if the parameter values are unrealistic. In order to explain why QE is satisfied in the glow curves of real materials and why their glow peaks are properly shaped like those of FO, it is important to know what are plausible values of parameters in real materials. This question is discussed in detail later in Chap. 5. It may suffice here to state that the defect center concentration of the impurity doped phosphors likes of which could be attributed to the traps and RC may lie in the range 10^{17}–10^{19} cm^{-3} and the relaxation probabilities 10^{-5}–10^{-9} cm^3 s^{-1}.

To the knowledge of this author, there are no other studies in the published literature attempting to provide an explanation for the prevalence of FO kinetics in the TL of the real materials. In fact the TL workers, in general the experimentalists, appear to be unmindful of the fact that the TL glow curves of real materials invariably display the characteristic property of FO kinetics, which is reflected by a stable value of T_m when the dose is changed. No change in T_m with change in dose is a property exclusively belonging to the FO kinetics of TL. Notwithstanding, as a general practice the experimentalists continue to analyze their glow curves using the GO kinetics equation or simply using the shape factor.

References

1. C. Kittel, *Introduction to Solid State Physics* (Wiley, New York, 1996)
2. R.H. Bube, *Photoconductivity of Solids* (Wiley, New York, 1960), p. 50
3. A.C. Lewandowski, S.W.S. McKeever, Phys. Rev. **B43**, 8163 (1991)
4. J.T. Randall, M.H.F. Wilkins, Proc. Roy. Soc. (London) series **A184**, 365 (1945)
5. M. Kumar, G. Chourasiya, R.K. Kher, B.C. Bhatt, C.M. Sunta, Indian J. Pure Appl. Phys. **47**, 402 (2009)
6. G.F.J. Garlick, A.F. Gibson, Proc. Phys. Soc. (London) **62**, 574 (1948)
7. R. Chen, J. Electrochem. Soc. (Solid State Sci.) **116**, 1254–1257 (1969)
8. C.E. May, J.A. Partridge, J. Chem. Phys. **40**, 1401 (1964)
9. C.M. Sunta, B.C. Bhatt, P.S. Page, in *Proceedings of the Third International Conference on Luminescence and its Applications*, Luminescence Society India, pp. 30–35, McMillan India Ltd. (2008)
10. M. Rasheedy, J. Phys, Condens. Matter **5**, 633 (1993)
11. C.M. Sunta, W.E.F. Ayta, R.N. Kulkarni, R. Chen, S. Watanabe, Radiat. Prot. Dosim. **71**(2), 93–97 (1997)
12. Y. Kirsh, Phys. Status Solidi (a) **129**, 15 (1992)
13. R. Chen, S.W.S. McKeever, *Theory of Thermoluminescence and Related Phenomena* (World Scientific, Singapore, 1997), p. 34
14. N. Takeuchi, K. Inabe, H. Nanto, Solid State Commun. **17**, 1267–1269 (1975)
15. A. Halperin, A.A. Braner, Phys. Rev. **117**(2), 408 (1960)
16. G.A. Dussel, R.H. Bube, Phys. Rev. **155**, 764 (1967)
17. I.J. Saunders, J. Phys. C Solid State Phys. **2**, 218 (1969)
18. C.M. Sunta, W.E.F. Ayta, R.N. Kulkarni, T.M. Piters, S. Watanabe, J. Phys. D Appl. Phys. **30**, 1234 (1978)
19. C.M. Sunta, W.E.F. Ayta, R.N. Kulkarni, T.M. Piters, S. Watanabe, J. Phys. D Appl. Phys. **30**, 1234–1242 (1997)
20. S.V. Moharil, S.P. Kathuria, J. Phys. D Appl. Phys. **16**, 425 (1983)
21. A. Opanowicz, Phys. Status Solidi (a) **116**, 343 (1989)
22. C.M. Sunta, W.E.F. Ayta, J.F.D. Chubaci, S. Watanabe, J. Phys. D Appl. Phys. **38**, 95–102 (2005)
23. C.M. Sunta, R.N. Kulkarni, T.M. Piters, W.E.F. Ayta, S. Watanabe, J. Phys. D Appl. Phys. **31**, 2074–2081 (1998)

24. R. Visocekas, La Luminescence de la calcite après irradiation cathodique. TL et luminescence par effettunel, Ph.D. Thesis, UniversitePiere et Marie Curie, Paris, 1978
25. R. Chen, N. Kristianpoller, Z. Davidson, R. Visocekas, J. Lum. **23**(3, 4), 293 (1981)
26. D. Shenkar, R. Chen, J. Comput. Phys. **10**, 272 (1972)
27. D. Yossian, Y.S. Horowitz, Radiat. Measur. **27**, 465 (1979)
28. R. Chen, V. Pagonis, Nucl. Instrum. Meas. Phys. Res. B **312**, 60–69 (2013)
29. C.M. Sunta, W.E.F. Ayta, J.F.D. Chubaci, S. Watanabe, J. Phys. D Appl. Phys. **34**, 3285 (2001)
30. C.M. Sunta, Phys. Status Solidi (a) **53**, 127 (1979)
31. C.M. Sunta, Radiat. Prot. Dosim. **8**(1/2), 25–44 (1984)
32. C.M. Sunta, W.E.F. Ayta, J.F.D. Chubaci, S. Watanabe, J. Phys. D Appl. Phys. **34**, 2690–2698 (2001)
33. C.M. Sunta, M. Kumar, R.K. Kher, B.C. Bhatt, J. Phys. D Appl. Phys. **39**, 4557–4562 (2006)
34. B. Yang, P.D. Townsend, A.P. Rowlands, Phys. Rev. **B57**, 178 (1998)
35. P.D. Townsend, A.P. Rowlands, Radiat. Prot. Dosimi. **84**, 7 (1999)
36. D. Brind, P. Jaconi, M. Bemabdsselam, D. Lapraz, P.W. May, C.A. Rege, Diamond Rela. Mater. **9**, 1245 (2000)
37. E.G. Yukihara, S.W.S. McKeever, E. Okuno, E.M. Yoshimura, Radiat. Prot. Dosim. **100**, 361 (2002)
38. C.M. Sunta, Phys. Status Solidi **37**, K81 (1970)
39. A. Mandowski, J. Phys. D Appl. Phys. **38**, 17 (2005)
40. M. Kumar, R.K. Kher, C.M. Sunta, J. Phys. D Appl. Phys. **39**, 2670 (2006)
41. C.M. Sunta, E.M. Yoshimura, E. Okuno, J. Phys. D Appl. Phys. **27**, 852 (1994)
42. P. Kelly, M.J. Laubitz, P. Braunlich, Phys. Rev. **B4**, 1960 (1971)
43. S.W.S. Mckeever, A.C. Lewandowski, B.G. Markey, Radiat. Prot. Dosim. **47**, 9 (1993)
44. A.C. Lewandowki, B.G. Markey, S.W.S. Mckeever, Phys. Rev. **B49**, 8029 (1994)
45. C.M. Sunta, W.E.F. Ayta, J.F.D. Chubaci, S. Watanabe, J. Phys. D Appl. Phys. **34**, 3285 (2001)
46. R. Chen, V. Pagonis, *Thermally and Optically Stimulated Luminescence: A Simulation Approach*, 1st edn. (Wiley, Chichester, 2011)
47. R. Chen, J. Mater, Sci. **11**, 1521 (1976)
48. R.K. Bull, J. Phys. D Appl. Phys. **22**, 1375 (1989)
49. S. Watanabe, Research Report to FAPESP, University of Sao Paulo Brazil (Institute of Physics, Sao Paulo, 2004)
50. J. Azorin, A. Gutierrez, Nucl. Tracks **11**, 167 (1986)
51. Y. Kirsh, P.D. Towsend, S. Shoval, Nucl. Tracks Radiat. Meas. **13**, 1509 (1987)
52. D. Dorendra Singh, S. Ingatomi, J. Phys. D Appl. Phys. **28**, 1509 (1995)
53. S.H. Tatumi, L.R. Batista, S. Watanabe, M. Matsuoka, Nucl. Instrum. Methods Phys. Res. A **280**, 510 (1989)
54. J. Prokein, G.A. Wagner, Radiat. Meas. **23**, 85 (1994)
55. Y. Kirsh, J.E. Townsend, P.D. Townsend, Phys. Status Solidi A **114**, 739 (1989)
56. M.A. EL-Kolaly, S.M.D. Rao, K.S.V. Nambi, A.K. Ganguly, Pramana **14**, 165 (1980)
57. M.S. Akselrod, E.A. Gorelova, Nucl. Tracks Radiat. Meas. **21**(1), 143 (1993)
58. C.M. Sunta, W.E.F. Ayta, R.N. Kulkarni, J.F.D. Chubaci, S. Watanabe, J. Phys. D Appl. Phys. **32**, 717 (1999)

Chapter 4
Kinetics Analysis of TL Glow Curves

Abstract This chapter deals with the kinetics analysis of TL glow peaks. Five methods of glow peak analysis are summarized. The purpose of glow curve analysis essentially is to find the activation energy E and the kinetic order b of the given glow peak. It is shown that the glow peaks shift to higher temperatures and get widened as b increases and trap occupancy (dose) decreases. However at $b = 1$ the glow peaks stay at the same temperature and retain their fixed shape irrespective of the trap occupancy. This property is exclusive to the first-order (FO) kinetics. It is therefore suggested that the first step in glow peak analysis should be to test whether the given peak shifts with trap occupancy (dose). If it retains the fixed temperature T_m with dose change it should be assigned FO. Application of peak shape method in analyzing the experimental glow curves may face uncertainty in the shape due to the presence of weak satellites close to the main peak under study. Simulations show that a linear relation exists between E and T_m for a given s/β, where s is the frequency factor and β is the heating rate. This trend is observed also in experimental samples. This implies that the value of the frequency factor s for the different glow peaks of a given sample may be same. It is found that due to the lacunae of general order (GO) kinetics model the found value of E by the glow curve fitting method is subject to error if the best fitted value of b is found to be different from 1 and 2.

4.1 Basic Parameters of TL Glow Peaks

As discussed in Chap. 3, the basic parameters on which the characteristics of a TL glow peak and its stability depends are the thermal activation energy E and the frequency factor s. In addition, the order of kinetics (KO) b comes into play if the glow peak is analyzed by using the general order (GO) kinetics model. However, as discussed in Chap. 3, a value of b different from 1 may be easily ruled out if the temperature T_m of the glow peak does not shift with change in dose. One should, however, remember that it is not a nominal change that matters, but the change in

C. M. Sunta, *Unraveling Thermoluminescence*,
Springer Series in Materials Science 202,
DOI: 10.1007/978-81-322-1940-8_4,

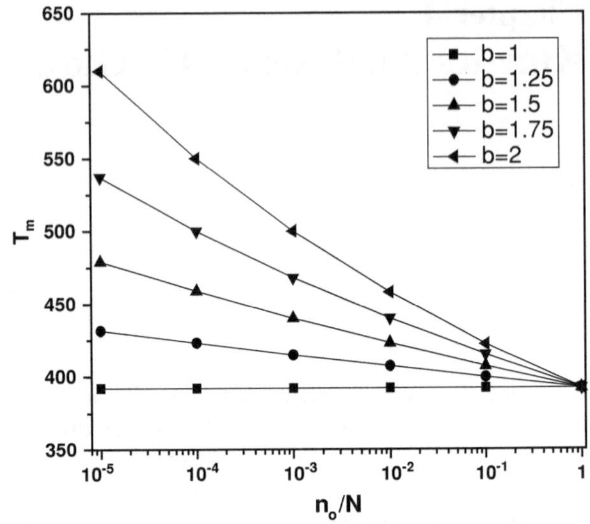

Fig. 4.1 T_m versus n_0/N (dose) plots for different values of b. Input parameters in simulation of glow curves are $E = 1$ eV, $s = 10^{12}$ s^{-1}, $N = 10^{17}$ cm^{-3} and $\beta = 2$ Ks^{-1}. The GO kinetics (3.21) is used to compute the glow curves of different b to find their respective T_m values

T_m on account of change in dose should be commensurate with the factor of dose change and the plausible b value. Figure 4.1 shows the change in T_m with change in b and with the trap occupancy n_0/N (assume the trap occupancy to be equitant to dose for a real sample). For example for $b = 2$ case, change in T_m per decade change in n_0/N is in the range of 30–60 K depending on the values of n_0/N. The value of n_0/N may be taken as the dose relative to the saturation dose (assuming that n_o changes linearly with dose). The shift in T_m for $b = 1.5$ case for similar changes in dose the shift in T_m ranges from 15 to 20 K. For a two-decade change in dose the shift in T_m becomes nearly double. Sometimes there may be closely spaced glow peaks and if some peak in such a glow curve grows faster than others with change in dose, the combined peak may appear shifted. Changes in glow curve profile are often seen at high dose exposures when the sample may suffer changes due to radiation damage. The experimenter has to make proper judgment to differentiate such observations from a genuine change in T_m due to non-first-order kinetics. The ambiguities because of such reasons may be largely avoided if the TL glow curve measurements are carried out at low dose exposures. It is advisable that the determination of TL glow peak parameters should be carried out at low dose exposures irrespective of the method of glow curve analysis used. Apart from avoiding some of the complications appearing at high doses as mentioned above, the change in T_m per decade change in dose also becomes progressively greater as dose (n_0/N) decreases. This becomes helpful to distinguish a genuine change of T_m due to non-first-order kinetics. Notwithstanding this discussion, it may be worthwhile to mention here that in our experience of measuring the glow curves of a variety of thermoluminescent materials we have not found any real material in which non-first-order glow peaks are observed by way of the peak shift with dose in a manner discussed above. Strong glow peaks followed by some weak intensity associated peaks may though produce a shape akin to non-first order. However, following Chen [1] earlier workers have invariably been basing their analyses on peak shape alone. For example, Takeuchi et al. [2], have reported that in MgO doped with iron, low concentration sample shows second-order ($b = 2$) behavior which with higher concentration produces a glow peak showing $b = 1.3$. The question why the computations of Chen did not show peak shift with change in b and with change in trap occupancy, is already discussed in Chap. 3 (see Sect. 3.3.3). In the following different methods of glow peak analysis are summarized.

4.2 First Approximation of *E*

An important datum of a glow curve is its T_m value. An expression to correlate T_m with E is obtained by setting $dI/dt = 0$ in the TL glow curve equation (see (3.12) in 3.3.1). For FO kinetics this yields

$$\frac{s}{\beta} = \frac{E}{kT_m^2} \exp\left(\frac{E}{kT_m}\right). \quad (4.1))$$

However this equation, being transcendental, cannot be solved in a straightforward manner to yield T_m as a function of E. Therefore, numerical calculations using (3.12) are carried out to find the value of T_m for various combinations of E and s/β. It is found that the values of T_m, as a function of E closely fall into a straight line for a given value of s/β (See Fig. 4.2). The actual relation between T_m and E for a given value of s/β may be fitted into

$$E = aT_m^x \quad (4.2)$$

where E is in eV, T_m in Kelvin and a and x are constants. The values of a and x are given in Table 4.1 for different input values of s/β. As may be seen in this table the value of x is nearly equal to 1. With these values of a and x, (4.2) is found to yield E values as a function of T_m within about 0.3 % error. In fact the change in x with change in s/β is so small that there appears to be a linear relation between E and T_m for a given value of s/β (see Fig. 4.2). The linear relation between E and T_m for simulated glow peaks of FO kinetics is shown also in Fig. 3.4 in Chap. 3. In Fig. 4.2 if we use $s/\beta = 10^{13}$, we get $E = \frac{T_m}{304}$. This may be considered as first approximation for the E value. The value of $s/\beta = 10^{13}$ would mean $s = 10^{13}\ \mathrm{s}^{-1}$ if the glow peak is recorded with $\beta = 1\ \mathrm{Ks}^{-1}$. The value of $10^{13}\ \mathrm{s}^{-1}$ may be a reasonable number to assume for s since this number is nearly equal to the Debye frequency. (See also Sect. 3.2.1). Urbach [3] arrived at the approximation $E = \frac{T_m}{500}$. It may be seen from Table 4.1 that this would nearly hold true for s/β value between 10^{11} and 10^{12}.

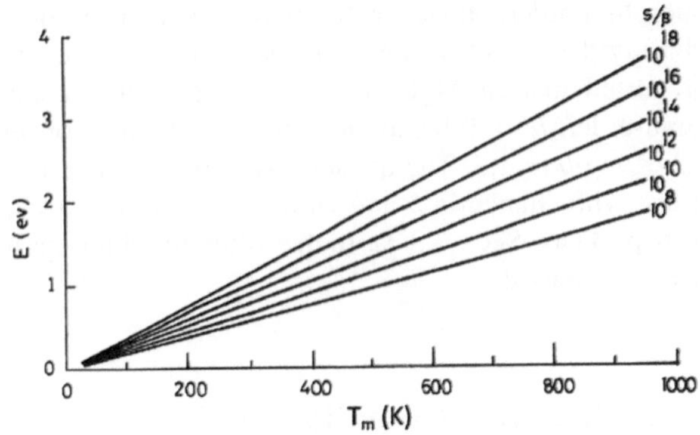

Fig. 4.2 E versus T_m for different s/b. These plots are drawn using the data of simulated FO kinetics glow peaks

Table 4.1 Constants a and x for different s/β values in (4.2) relating the glow peak temperature T with the activation energy E

S/β	a	x
10^8	1.58×10^{-3}	1.038
10^{10}	1.766×10^{-3}	1.038
10^{11}	1.953×10^{-3}	1.0350
10^{12}	2.141×10^{-3}	1.0325
10^{13}	2.329×10^{-3}	1.0303
10^{14}	2.519×10^{-3}	1.0284
10^{15}	2.709×10^{-3}	1.0267
10^{16}	2.9×10^{-3}	1.0252
10^{18}	3.283×10^{-3}	1.0226

4.3 Initial Rise Method

In the very initial part of a glow peak, when the TL intensity I has risen by no more than 15 % of the maximum intensity of the glow peak I_m, the depletion in the population of the trapped carriers is negligible. This means that the value of n may be considered to be unchanged from its initial value n_0 in this part of the glow peak. Thus, the temperature-dependent quantity within the square brackets in the glow curve (3.12), (3.15), and (3.17) of the FO , second order (SO), and GO kinetics becomes equal to 1. The relative intensity in the initial part of the glow curve may then be represented by

$$I = C\exp\left(-\frac{E}{kT}\right)'' \tag{4.3}$$

which is independent of the KO applicable to any given glow peak. A plot of ln I, versus $1/T$ called the Arrhenius plot produces a straight line with slope equal to $-E/k$ and intercept equal to ln C when the sample is heated to raise the temperature linearly with time (see Fig. 4.3). The value of C is equal to (sn_0), (sn_0^2/N), and $(s'n_0^b)$ for FO, SO, and GO kinetics, respectively. One may get the value of E from the

Fig. 4.3 ln TL intensity I versus $1/T$ plots for the initial rise parts of the simulated glow peaks of kinetics order *1* $b = 1$ and *2* $b = 1.5$. *Curves a, b* and *c* both in *1* and *2* are respectively are for $n_0/N = 1$, 10^{-2} and 10^{-4}. Input parameter valued in glow curve calculation are $E = 1$ eV, $s = 10^{12}$ s^{-1}, $N = 10^{17}$ cm^{-3}

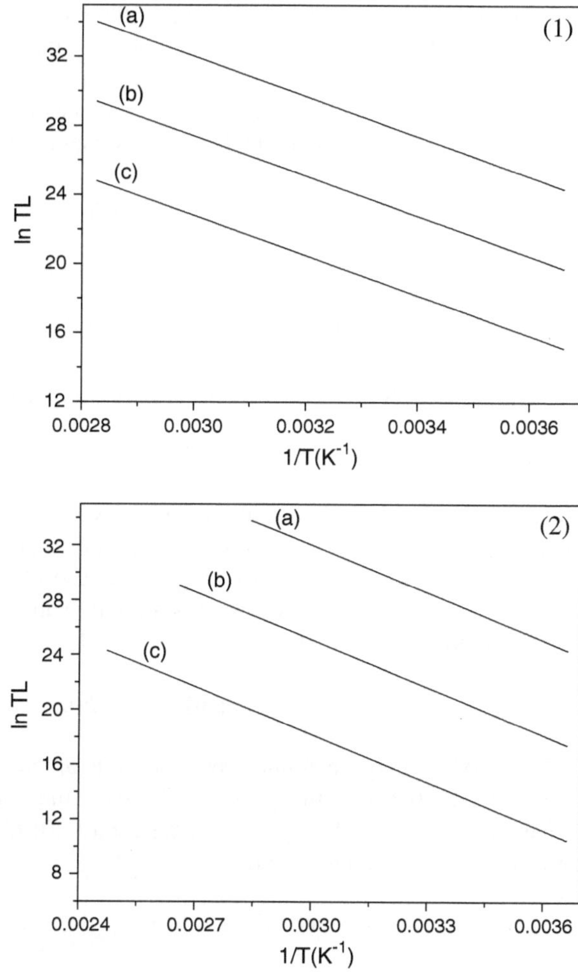

slope. This method as stated above is applicable to all single energy glow peaks irrespective of their kinetic order and was first suggested by Garlick and Gibson [4].

The precautions which are essential for using this method are: (1) the initial rise part of the glow peak under study should be free from any contribution from any nearby glow peaks and (2) there should be no lag in temperature between the heater plate and the sample and no temperature gradient across the sample thickness. This means that the sample layer should be very thin and it should be firmly deposited on the heater pan. Flow of dry nitrogen gas through the sample heating chamber also helps in producing better conduction between the sample and its heater base apart from avoiding minor spurious signal like triboluminescence. Use of slow heating rate is desirable to avoid temperature lag between the heater pan and the sample.

Rawat et al. [5] have shown that the initial rise method may also be used to find the kinetic order b. The suggested method uses the intercept values on the Y axis of the Arrhenius plots for two or more sample doses. Using the equation of GO kinetics (see (3.16) and (3.16a)) with $n = n_0$ (which is applicable for initial rise intensities) the Y axis intercept I_c of the Arrhenius plot is

$$I_c = \ln s' n_0^b. \tag{4.4}$$

If the trap occupancy (dose) is increased by a factor of x, the intercept becomes

$$I_{cx} = \ln s'(x\, n_0)^b = \ln (s' n_0^b) + b \ln (x). \tag{4.5}$$

Using the difference between these two intercept values, one gets

$$b = (I_{cx} - I_c)/\ln(x). \tag{4.6}$$

One may record initial rise at more than two doses to improve the confidence on the found value of b. Equation (4.6) is general, in the sense that it is applicable for any value of b. Rawat et al. also suggest the use of the intercept value to find the pre-exponential factor s'. As already stated the value of the intercept I_c for the GO kinetics is $\ln (s' n_o^b)$. This leads to

$$s' = \text{anti} \ln[I_c - b \ln(n_0)]. \tag{4.7}$$

Value of n_0 may be found from the area of the full glow peak provided the scaling factor between the area and the n_0 value could be known. The value of b may be found by utilizing (4.6). If we use (3.20) given in Chap. 3 for s', we can find the frequency factor s, viz

$$s = \text{anti} \ln \,[I_c - \ln(n_0) + (b - 1)\ln(n_0/N)]. \tag{4.8}$$

This expression is general, in the sense that it is applicable for all values of b. It gives the meaningful parameter s, whereas (4.7) gives the empirical constant s'. The value of (n_0/N), for use in (4.8) may be estimated by the ratio between total TL intensity (area of the glow peak) at the given dose (n_0) and at the saturation dose ($n_0 = N$).

4.3.1 Analysis of Multi-peak Glow Curves

The above discussion is based on a single isolated glow peak. However, in real materials a single peak glow curve is a rarity. The TL glow curves of most of the sample materials have multiple glow peaks. Further in some cases there are very closely spaced glow peaks, such that parts of individual peaks overlap on the adjoining peaks. In case the glow peaks are fairly well separated, for example, in a mineral CaF_2 sample (Sunta [6]). One may start with the lowest temperature glow peak and bleach it out after finishing measurement on this peak. Bleach out may be done by heating up to a temperature above that of the valley between the adjoining peaks. In this way measurements may be carried out one by one on individual glow peaks. Each time the sample is cooled down to room temperature and then the initial rise is recorded for the next glow peak. If the glow curve consists of overlapping glow peaks, for example in CaF_2: Dy and $CaSO_4$: Dy samples (see for example Sunta [6]), one may still use the initial rise method. One has, once again, to begin with the lowest temperature part of the glow curve. After recording the initial rise for this part, the sample is cooled and once again the initial rise is recorded. Thus, by repeating the recording of the initial rise and cooling cycles several times until the end of the glow curve, one may get a series of ln I versus $1/T$ straight line plots. All measurements are carried out with the same heating rate. One expects some of the consecutive plots of the initial rise curves to have similar slopes, meaning thereby that they belong to the same glow peak. We may get successive groups of this type. One may thus get an E value for each of the groups. The E value obtained from each ln I versus $1/T$ plot may be plotted against the midpoint temperature T of the corresponding initial rise curve. This has been named as fractional glow curve technique by Gobrecht and Hoffman [7]. Using this method ladder-like E versus T curves have been obtained for diamonds by Nahun and Halperin [8]. Kirsh et al. [9] have reported similar results for albite and microcline mineral samples [9]. Srivastava and Supe [10] have studied $CaSO_4$: Dy TL phosphor [10] by the same method. The result is shown in Fig. 4.4a. In this method, the E values obtained from groups of individual Arrhenius plots for a landing level of the ladder scatter around an average. The method thus yields approximate values of E. This procedure has shown that the E values of the glow peaks of a sample rise almost linearly with the peak temperature T_m in the multi-peak glow curve of a sample. Figure 4.4b shows the E versus T_m plots for the glow curves of albite, microcline, and $CaSO_4$: Dy samples. This result supports the theoretical one discussed in the beginning of the chapter showing that the E value is nearly directly proportional to T_m for any given s/β . It is also in agreement with the simulated glow peaks of different E values shown in Chap. 3, Fig. 3.4. The above result on the samples of albite, microcline, and $CaSO_4$: Dy implicitly demonstrates that for a given sample the s value is the same for its different glow peaks.

4.3.2 Sources of Error

Braunlich [11] has shown by simulation that the initial rise method fails in a system in which the retrapping probability far exceeds the recombination probability ($A_n \gg A_h$) and the traps are filled to saturation ($n_0 = N$). Sunta et al. [12] have later shown that the so-called failure of the initial rise method seen by

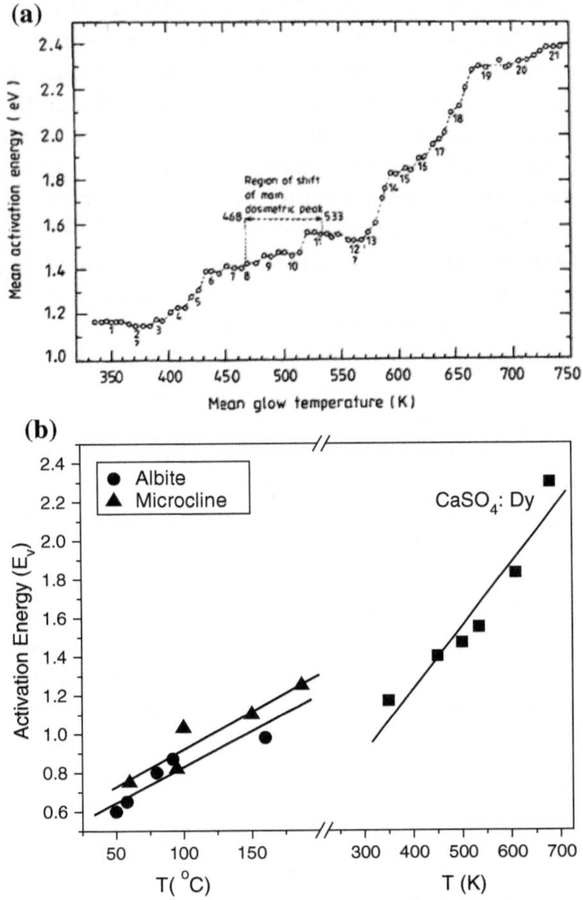

Fig. 4.4 **a** The profile of mean activation energy *E* versus temperature curve of fractional glow curve technique (FGT) for $CaSO_4$: Dy TL phosphor. The glow curve of $CaSO_4$: Dyshows distinct peaks at about 420, 510, 620 and 675 K with minor kinks in between them. The glow peak of 510 K being strongest is used in dosimetric applications. This peak seems to shift its mean position between 468 and 533 K. (after Srivastava and Supe [10, Fig. 2]. © IOP Publishing. Reproduced by permission. All rights reserved. **b** Mean activation energy*E* versus T_m for $CaSO_4$: Dy (data from **a** above) and for albite and microcline (data from Gobrecht and Hoffman [6], and Kirsh et al. [7]). The values of activation energy as well as those of the Temperature are approximate from the FGT method

Braunlich is due to the fact that in such cases the valid initial rise part in which the growth of the TL intensity is given by Boltzmann function *C* exp (−*E/kT*) lies below 1 % of the peak intensity (See curve 1 in Fig. 4.5), whereas Braunlich used the rule of thumb which allows the region up to 15 % of the peak intensity as the valid initial rise part. Sunta et al. [10] show that even in the high retrapping cases, the valid initial rise part comes up to 15 % of peak as usual if the trap occupancy is kept at less than 1 % of the saturation (Compare curves 1 and 2 in Fig. 4.5). This means that the peak analysis needs to be carried out by using the low dose samples. In fact in any of the methods of glow peak analysis, as stated in the beginning of this chapter, it is advisable to avoid using near-saturation doses given to the sample.

A source of error in the initial rise method may sometimes be thermal quenching effect if this effect is present in the temperature span of the glow curve. Sunta and Bapat [13] have estimated the effect of thermal quenching on *E* values of $CaSO_4$: Ce and $CaSO_4$: Eu. They recorded the X-ray-excited radioluminescence

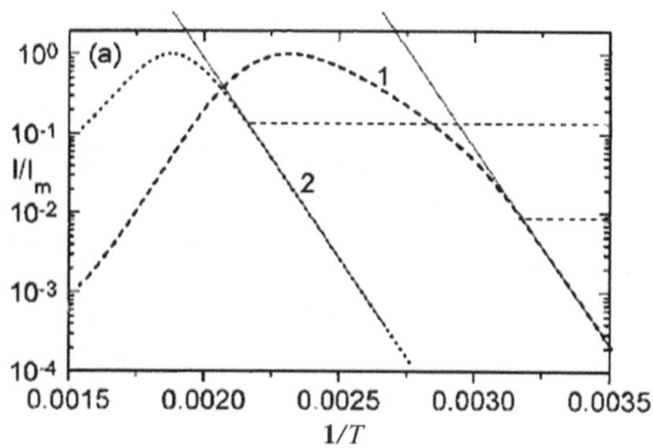

Fig. 4.5 I/I_m versus $1/T$ plots (Arrhenius plots) for simulated glow peaks of OTOR model. *Curve1* is for the saturated ($n_0/N = 1$) glow peak and *curve2* is for the unsaturated glow peak ($n_0/N = 10^{-2}$). The input parameter values for these glow curves are $E = 1$ eV, $s = 10^{12}$ s^{-1}, $N = 10^{10}$ cm^{-3}, $A_n = 10^{-7}$ cm^3 s^{-1}, $A_h = 10^{-9}$ cm^3 s^{-1} and $\beta = 1$ Ks^{-1}(after Sunta et al. [12])

(RL) and the photo-excited luminescence (PL) intensities as a function of sample temperature. Since the luminescence emission spectra were seen to be same in RL, PL, and TL, the thermal quenching effect in TL may legitimately be considered to be the same as in RL and PL. Applying the correction for thermal quenching, the E values by initial rise method were found to change by about 5 % in $CaSO_4$: Ce and by about 9 % in $CaSO_4$: Eu.

4.4 Various Heating Rates Method

This method is based on the property of the glow peaks that the peaks shift with change in heating rate β. The temperature T_m of maximum intensity for an FO glow peak can be found by equating the derivative of (3.12) (see Sect. 3.3.1) to zero. This yields

$$\frac{\beta E}{kT_m^2} = s\exp\left(-\frac{E}{kT_m}\right). \tag{4.9}$$

Rewriting this equation as

$$\beta = \frac{skT_m^2}{E}\exp\left(-\frac{E}{kT_m}\right) \tag{4.10}$$

we see that when β is increased, the function on the right-hand side of this equation also has to increase, which implies T_m would increase with increase in β. Since E and s are constant for a given glow peak, one gets increase in T_m with increase

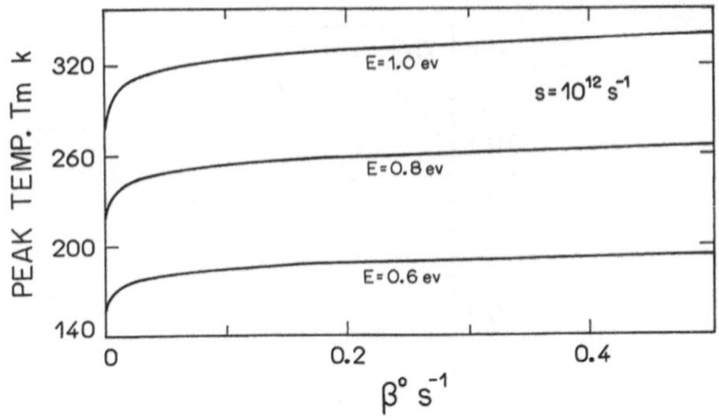

Fig. 4.6 T_m versus β plots for different E. Value of s used is 10^{12} s^{-1}

in β. Figure 4.6 shows T_m as a function of β for given values of E and s. Rewriting (4.9) for heating rates β_1 and β_2 and corresponding peak temperatures T_{m1} and T_{m2} and dividing one by other and rearranging the resulting expression, we get

$$E = \frac{kT_{m1}T_{m2}}{(T_{m1} - T_{m2})} \ln\left[\frac{\beta_1}{\beta_2}\left(\frac{T_{m2}}{T_{m1}}\right)^2\right]. \tag{4.11}$$

The selected heating rates β_1 and β_2 should be such that the peak temperatures T_{m1} and T_{m2} are well separated from each other. If T_m can be measured within an accuracy of 1 %, the method would yield the value of E within an accuracy of about 5 %. Having determined E, the value of s may be obtained by substituting the values of E and T_m in (4.9).

4.4.1 Hoogenstraaten's Method

Hoogenstraaten [14] suggested another way of using (4.9) for determining E and s. Rearranging (4.9) and taking natural logarithm one gets the following expression:

$$\ln\left(\frac{T_m^2}{\beta}\right) = \frac{E}{k}\left(\frac{1}{T_m}\right) - \ln\left(\frac{sk}{E}\right) \tag{4.12}$$

This method requires several measurements of the glow peak with different β. A plot of $\ln(T_m^2/\beta)$ versus $1/T_m$ would produce a straight line. The slope of the line (E/k) and its intercept $\ln(sk/E)$ are used to calculate the values of E and s, respectively. The value of s can thus be determined in two ways, namely by substituting the value of E in (4.9) or in the measured intercept $\ln(sk/E)$ of the straight line. The intercept is obtained by extrapolating the $\ln(T_m^2/\beta)$ versus $1/T_m$ straight line to $1/T_m \rightarrow 0$.

The various heating rates methods, though is based on FO kinetics, it has been found that it turns out to be a very good approximation even for any non-first-order kinetics. Singh et al. [15] and Chen and Winer [16] have analyzed this subject in detail. Chen and Winer found the method successful for computer generated non-first-order TL peaks as well as for the experimentally measured glow peaks of ZnS: Eu^{3+}.

4.4.2 *Merits, Demerits, and Precautions*

The various heating rates method has the merit that it is easy to use, since it requires to locate only the T_m for any selected heating rates β. It may avoid the difficulty that may arise due to the presence of weak satellite peaks. However, if overlapping glow peaks produce a maximum which is the combined result of more than one glow peak, the situation becomes complex. In such a case judicious thermal cleaning may be helpful to distinguish the single peak maximum. The choice of the heating rate is an important aspect for the various heating rates method. Too low a heating rate produces broad glow peaks, in which precisely locating the T_m value may pose some uncertainty. Slow heating rates though have an advantage that the change in T_m with change in heating rate is more at slower heating rates than that from the similar change in heating rate when the heating rates are fast (Fig. 4.6). On the other hand, fast heating rate produces a sharp peak which makes it easier to locate the T_m, but too fast a heating rate may lead to a lag between the temperature of the sample and the heater pan whose temperature is measured by the thermocouple. Further, a temperature gradient may exist within the sample thickness in fast heating. Use of very thin sample layer is therefore strongly advised in addition to judicious selection of heating rates.

4.5 Isothermal Decay Method

This method was first given by Randall and Wilkins (RW) [17]. If we assume an isolated FO kinetics glow peak, its E value may be found by simply plotting its isothermal decay at different temperatures. The TL intensity $I(t)$ at various times t at any given temperature T is given by

$$I(t) = I_0 \exp\left(-P(T)t\right) \tag{4.13}$$

where I_0 is the intensity at $t = 0$ and $p(T)$ is the intensity decay rate constant at temperature T and is given by where

$$p(T) = s \exp\left(-E/kT\right). \tag{4.14}$$

One may plot ln $[I(t)/I_0]$ versus t at two temperatures, say T_1 and T_2 and obtain the corresponding $p(T_1)$ and $p(T_2)$ values from the respective slopes of the obtained straight lines. The ratio $p(T_1)/p(T_2)$ leads to the value of E as

$$E = \frac{kT_1T_2}{(T_1 - T_2)}\ln\left[\frac{p(T_1)}{p(T_2)}\right] \tag{4.15}$$

The value of s may be found by substituting the value of measurement temperature T, the measured values of E and $p(T)$ in the above stated expression of $p(T)$. To improve the confidence in the obtained values of E and s, one may record the isothermal decay of the given sample at several temperatures, to obtain corresponding $p(T)$ values at the selected decay temperatures. One may now plot ln $p(T)$ values as a function of $1/T$. This would produce a straight line with slope equal to $(-E/k)$ and intercept equal to ln s. In this way both E and s may be obtained. An alternative method is to measure the glow peak (area or height) after maintaining the sample at a steady temperature for a certain period. The experiment is repeated for various periods of decay to obtain the isothermal decay curve for the used temperature. The whole experiment may be repeated at different temperatures.

If the glow peak is judged to be of non-first-order based on a consistent peak shift with change in dose (see Fig. 4.1), one may still use the isothermal decay method. The decay pattern obviously will be nonexponential. The method may be understood from the explanation as given below. Using the GO kinetics expression (3.16) (see Sect. 3.3.3), one may write

$$-dn/n^b = s'\exp(-E/kT)\mathrm{d}t.$$

Integrating this equation we have

$$n = [n_0^{1-b} + (b-1)s'\exp(-E/kT)t]^{1/1-b}. \tag{4.16}$$

Using (4.16) in (3.16) we have for the isothermal decay

$$I = [n_0^{1-b} + (b-1)s'\exp(-E/kT)t]^{b/1-b}s'\exp(-E/kT). \tag{4.17}$$

Normalizing it by $I_0(= n_0^b/s'\exp(-E/kT))$, we have

$$I/I_0 = [1 + n_0^{b-1}(b-1)s'\exp(-E/kT)t]^b/^{1-b}.$$

This leads to

$$(I/I_0)^{(1-b)/b} = 1 + n_0^{b-1}(b-1)s'\exp(-E/kT)t. \tag{4.18}$$

This shows that a plot of $(I/I_0)^{(1-b)/b}$ versus t at the given temperature T will produce a straight line with slope S given by

$$S = n_0^{b-1}(b-1)s'\exp(-E/kT). \tag{4.19}$$

However, it is to be noted that a straight line would be produced for $(I/I_0)^{(1-b)/b}$ versus t in (4.18) only when the correct value of b is inserted. To find the correct value isothermal decay data should be plotted with different values of b. That value which produces a straight line is the actual value of b [18]. Recording of isothermal decay at various T for a given n_0 (sample with same dose) would produce different values of S. A plot of ln(S) against $1/T$ now gives the value of E. We thus have both E and b values from the isothermal decay for the GO kinetics case.

An important characteristic to be noted, in this case is that the isothermal decay rate is dependent on n_0 (see (4.18) and (4.19)). As a consequence of this the obtained values of E and b will depend on n_0 (i.e., dose given to the sample). This shows the fundamental lacuna in the GO kinetics expression namely that no fixed values of E and b may be assigned to a given glow peak if the found value of b is different from 1. The lacunae in GO kinetics are already discussed in Chap. 3.

A potential source of error specific to isothermal decay method is the possibility of change in the constitution of the sample due to maintaining it at a higher temperature during recording of isothermal decay. In some materials, particularly in alkali halides, defect clustering, and precipitation of dopant impurities is seen. This may manifest in the form of midway distortions in the isothermal decay graph.

4.6 Methods Based on the Shape of the Glow Peak

Chen and McKeever [19] have extensively reviewed the work of various authors who have contributed to the development of the peak shape methods. In the work of these various authors expressions for the trap depth E are developed using the approximation that the glow peak may be described as consisting of two right-angled triangles of same height. The methods essentially use only three temperature points on the glow curve, namely T_m, T_1 , and T_2 where T_m is the temperature of maximum intensity and T_1 and T_2 , respectively, are the temperatures at half the maximum intensity on the rising and the falling side of the peak. The half width of the full glow peak is given by $\omega = T_2 - T_1$ and those of the two right-angled triangles as $\tau = T_m - T_1$ and $\delta = T_2 - T_m$. These half widths are assumed to be proportional to the areas of the respective triangles, which in its turn means the number of traps emptied to produce the TL in the full glow curve, in the first half and in the second half, respectively (see Fig. 4.7). Various authors developed expressions for E utilizing either τ, δ, or ω and assuming either first- or second-order kinetics. These expressions have been summed up into a single expression (Chen and McKeever [19], Chen [1]). It is given as

$$E_\alpha = c_\alpha\left(kT_m^2/\alpha\right) - b_\alpha(2kT_m) \tag{4.20}$$

where c_α and b_α are constants in which α stands for τ, δ, or ω. Table 4.2 shows the values of the constants c_α and b_α for first- and second-order kinetics. For distinction between first- and second-order kinetics Halperin and Braner [20]

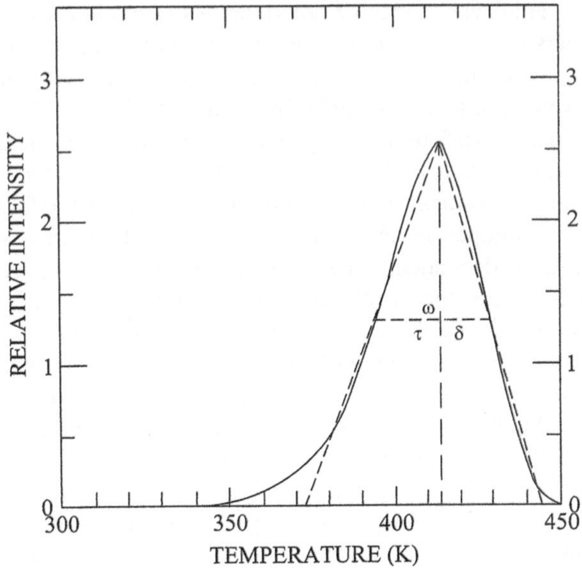

Fig. 4.7 A TL glow curve approximated to two *right* angled triangles. The half width is represented by ω, the first half of ω is τ and second half is δ

Table 4.2 The values of coefficients c_α and b_α in Eq. (4.20) for three different shape methods for calculating the E value of first and second order kinetics glow peaks

	First order (FO)			Second order (SO)		
	τ	δ	ω	τ	δ	ω
c_α	1.51	0.976	2.52	1.81	1.71	3.54
b_α	$1.58 + a/2$	$a/2$	$1 + a/2$	$2 + a/2$	$a/2$	$1 + a/2$

a is the power of T if the frequency factor changes with T, usually $-2 \le a \le 2$. If s is assumed to be temperature independent (i.e $a = 0$), the b_α value would be simplified in the table

suggested the use of the symmetry factor μ of the glow peak which is defined as $\mu = \delta/\omega$. An FO peak is characterized by $\mu = 0.42$ and a second-order peak by $\mu = 0.52$. For the general case in which KO is not necessarily equal to 1 or 2, Chen [1] investigated the correlation between μ and the three parameters, namely the activation energy E, the pre-exponential factor ($s'' = s'n_0^{b-1}$), and the kinetics order b. Chen used the glow curve Eq. (3.18) (refer Sect. 3.3.3) of the GO kinetics model for this study and found that μ is strongly dependent on b and practically independent of E and s''. The computations covered wide range of input values namely $0.7 \le b \le 2.5$, $0.1 \le E \le 1.6$ eV, and $10^5 \le s'' \le 10^{19}\,\mathrm{s}^{-1}$. The result is shown in Fig. 4.8. It may be seen that the deviations in μ values for various E and s'' are within about ±7 %. Figure 4.8 may be used to find the value of b from the measured value of μ in an experimentally measured glow peak. To find the values of c_α and b_α in (4.20) for $b \ne 1$ and $\ne 2$, Chen [1] used the measured μ value as an interpolation parameter along with known values of $\mu = 0.42$ and $\mu = 0.52$ for $b = 1$ and $b = 2$ respectively. The expressions for c_α and b_α thus found are

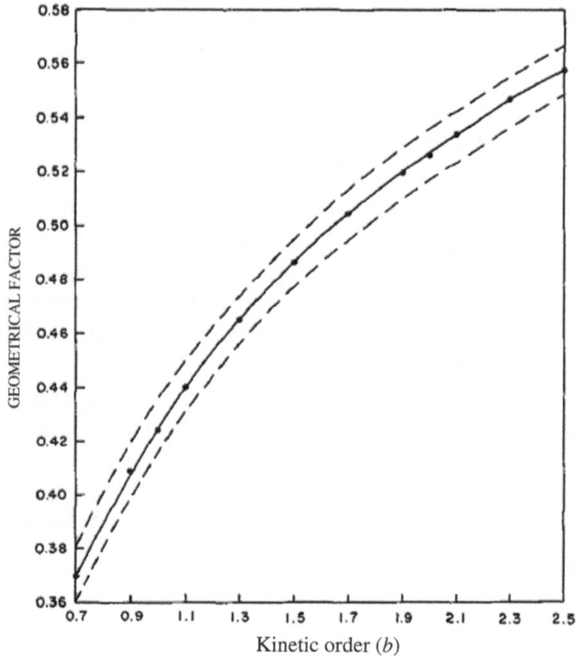

Fig. 4.8 Geometrical shape factor of a general order kinetics glow curve. (after Chen [1, Fig. 1]. Reproduced by permission of ECS—The Electrochemical Society

$$c_\tau = 1.51 + 3(\mu - 0.42); b_\tau = 1.58 + 4.2(\mu - 0.42) \tag{4.21}$$

$$c_\delta = 0.976 + 7.3(\mu - 0.42); \; b_\delta = 0 \tag{4.22}$$

$$c_\omega = 2.52 + 10.2(\mu - 0.42); \; b_\omega = 1. \tag{4.23}$$

The c_α and b_α values found from these equations may now be used in (4.20) to get the value of E.

As an extension of the above method, Singh et al. [21] have suggested to use the δ/ω values at 2/3 and 4/5 of the peak intensity which may avoid possible interference from any weak satellites, since the weak intensity satellite peaks are likely to interfere more at the lower parts of the glow peak than at the upper parts.

Table 4.2 below shows the values of coefficients c_α and b_α in (4.20) for three different shape methods for calculating the E value of first- and second-order kinetics glow peaks. a is the power of T if the frequency factor changes with T, usually $-2 \leq a \leq 2$. If s is assumed to be temperature independent (i.e., $a = 0$), the b_α value would be simplified as may be seen in the table.

4.6.1 Precautions and Sources of Error

As stated before in this method also the purity of the glow peak is an essential requirement for the accuracy of the results. In addition correct measurement of

T_1, T_2, and T_m is essential to get the values of τ, δ, or ω. Chen [1] found that the method using τ gives more error in comparison to the methods using δ or ω, Chen has suggested that it is best to use c_ω and b_ω in (4.20) for calculating E. Apparently, the reason for the greater error in the method using τ could be that the fractional area left outside the first half of the triangle is a larger fraction of that triangle than the fractional area left out of the full triangle in the case of using the ω method. In any case, the peak shape method is an approximate method in the sense that the value of the peak shape factor μ is not totally independent of the E and s values. It varies up to ± 7 % from its mean value for different E and s. Shenker and Chen [22] used this method for the simulated TL peaks. They report that the errors in the computed values of E are within 5 % of the input values used in the simulations.

It needs to be reminded here that before applying the shape method one should test the glow peak for shift in T_m with change in dose. In case there is no consistent shift, as stated in the beginning of this chapter, one should assume the glow peaks to belong to the FO kinetics. Difference in shape in comparison with that of FO invariably arises because of associated weak intensity components with the main glow peak.

4.7 Curve Fitting Method

In the preceding paragraphs, we have discussed the methods that can analyze only one peak at a time and the methods require that the peak should be pure or it may have been purified by a procedure called thermal cleaning, which means partial readout of the glow curve to remove the initial part if any low intensity peaks are suspected to be present in that part. The curve fitting technique on the other hand is capable of analyzing all the glow peaks in a multi-peak glow curve of a real sample. The popularity of the curve fitting technique has stemmed from the advancements in computational techniques, in which large number of numerical data points may be handled at one go. The basic approach for glow curve fitting involves the following: (1) choosing the appropriate equation to fit the data. For example, one may choose the Randall-Wilkins (RW) equation of the glow peaks (see (3.12) in Sect. 3.3.1). This would implicitly mean that all the glow peaks in the multi-peak glow curve belong to FO kinetics ($b = 1$). One may then carry out a three parameter fit namely using E, s, and n_0 as fitting parameters. The fitting parameter n_o takes care of differences in intensities of different peaks in a multi-peak glow curve. The glow curve fitting by most authors, however, is carried out using the GO kinetics equation of the glow peak (see (3.17) in Sect. 3.3.3). This requires a four parameter fit, namely E , s', b , and n_0 as fitting parameters. In place of s' one may use s'', where $s'' = s\, n_0^{b-1}$ which was introduced by Chen [1]. Alternatively if one uses (3.21), it would mean a five parameter fit namely E, s, b, n_0 , and N. The relevance of the values of n_0 and N may be nothing more than the fitting parameters and not the real numbers applicable to the given sample. (2) A clear judgment is needed about the number of glow peaks in the multi-peak glow curve under analysis. If the number of glow peaks is g, one would have to fit 4 g parameters if (3.17) is used and 5 g parameters if (3.21) is used. It is, however, seldom easy to judge the number of peaks in a multi-peak glow curve. As discussed later in this chapter, the uncertainty regarding the number of peaks may lead to dubious results. The analysis usually proceeds by using a simple linear superposition of the selected equation to cover the g overlapping glow peaks. Each of the 4 or 5 g parameters (depending on the equation used) is allowed to vary independently until a 'best fit' is arrived. The process of arriving at the "best-fit" involves the minimization of a pre-selected error function which gives the overall mismatch between the experimental data points and the fitted points obtained in the linear superposition of the assumed function. The error function may be the so-called "figure of merit" (FOM) or a simple "chi-square." The "best fit" is taken as that at which the error function converges to a minimum. Notwithstanding the elaborate nature of the procedure, one may not be able to attain certainty in locating the minimum. It is therefore advisable to first make estimates of parameters by using the analytical techniques described above in this chapter. These may then be used as starting values in the fitting process. As a first step, whether the glow peaks are FO or non-first-order kinetics may be easily found by testing the dose dependence of T_m of the glow peaks. If there is no shift in T_m by changing the dose by, say one or two orders of magnitude, one should take $b = 1$. One may then use the RW equation and proceed with a three parameter (E, s, and n_o) fit. The function FOM is defined as under:

$$\mathrm{FOM} = \left| \sum (y_i - y_f) \Big/ \sum y_i \right| \times 100 \tag{4.24}$$

where y_i are the measured values of intensity I at various T in the experimental data and y_f are the corresponding best fitted values.

4.7.1 Limitations of Peak Shape and Curve Fitting Methods

Although these two methods are most popular among the TL workers for the TL glow curve analysis, it has been found that under certain parametric conditions these methods may fail. This happens in particular when the retrapping probability coefficient A_n is much greater than the recombination probability coefficient A_h and the active traps are filled to saturation ($n_0 \approx N$). This kind of case has been used by Braunlich [11] to show the limitation in the use of initial rise method and is already discussed in Sect. 4.3.2. In the following, we discuss a similar case used in peak shape and peak fitting method. Sunta et al. [12] simulated the TL glow peaks using the rate equation of the OTOR model (see Sect. (3.3.4)) and using the input parameter values $E = 1$ eV, $s = 10^{12}$ s^{-1}, $N = 10^{10}$ cm^{-3}, and $A_n = 10^{-7}$ cm^3

Table 4.3 Change in found values of E with change in n_0/N (dose) (input = 1 eV) at high values of A_n/A_h

Input parameter		Found values of E (eV)			
A_n/A_h	n_0/N	Peak shape methods $E_\tau E_\delta E_w$			Peak fitting method E
10^3	1	0.466	0.556	0.512	0.511
10^2	1	0.470	0.549	0.510	0.487
10^2	10^{-1}	0.986	0.994	0.994	0.987
10^2	10^{-2}	0.998	0.994	0.998	1.047

It may be seen that at $n_0/N < 10^{-1}$, the found E values are nearly equal to the input value. (Values taken from Sunta et al. [12])

s^{-1} and various values of n_0 and A_h such that n_0/N varies from 1 to 10^{-2} and A_n/A_h varies from 10^{-2} to 10^2. They back-calculated the E value of these simulated glow curves using the peak shape as well as the peak fitting method. The results show that when the traps are saturated ($n_0/N = 1$) and the $A_n/A_h \geq 1$, the found value of E becomes progressively smaller than its input value as A_n/A_h increases. At $A_n/A_h = 10^2$, the found E value is seen to be about 50 % of the input value both in peak shape as well as peak fitting method. However, when the filled trap population is well below saturation ($n_0/N \leq 10^{-1}$) the E value found is seen to become nearly equal to the input value even when the value of A_n/A_h is as high as 10^2. A summary of results from Sunta et al. [12] is shown in Table 4.3. It may be recalled that similar anomaly is encountered also in the case of the initial rise method at $n_0 \approx N$ when $A_n/A_h \gg 1$. In measuring the E of real materials it may be easily avoided by using the low dose samples (see Sect. 4.3.2). The important lesson from this exercise is that, whether it is for using the peak fitting and the peak shape methods, or for using the initial rise method, the sample dose should be one to two orders of magnitude below the saturation dose. It may be interesting to note that the difference in the E values obtained by the peak fitting and the peak shape methods (particularly the method based on E_ω) is not very significant even though the former method uses the entire glow curve and the latter uses only the three points in the glow peak. However, the peak fitting method has the advantage that it can analyze all the glow peaks in a multi-peak glow curve, whereas the peak shape method can analyze only one peak at a time. It may also be noted here that the failure of the initial rise method under the saturation dose and $A_n/A_h \gg 1$ conditions described earlier, is due to the inappropriate use of the 'rule of thumb' method, which allows the intensity up to 15 % of the peak value as the initial rise part, whereas under the special conditions mentioned above the valid initial rise (pure exponential rise) region is only up to 1 % of the peak intensity (see Fig. 4.5). The initial rise method may therefore still be used even under the above-mentioned special conditions if the TL phosphor is sensitive enough, since in the case of sensitive TL phosphors the glow peak may easily be recorded up to two to three decades below the peak intensity. In contrast, in the case of the peak fitting and the peak shape methods, the reason for their failure is inherent in these methods themselves.

4.8 General Precautions and Uncertainties in Glow Curve Analysis

(1) *Temperature measurement*—As stated before in glow curve recording accurate measurement of temperature is most important. Necessary precautions to ensure this are already mentioned in Sect. 4.3. If the sample is in the form of fine powder it should be firmly deposited on the heater plate as described for fine grain TL dating technique (Zimmermann [23]). This involved technique may, however, not be always necessary. A simper method may be to smear a very light layer of silicone resin on the heater plate before placing the sample on it. The sample may be in the form of thin polished plate of approximately 0.5 mm or less thickness. If the sample is in the form of fine powder one may use ≤ 5 mg of the powder which may be uniformly spread on the heater plate over about 1 cm^2 area uniformly.

(2) *Purity of the glow peak*—The methods described above assume that the glow peaks to be analyzed are free from any satellites associated with the glow peak under study. As described earlier satellites on the lower temperature side may be bleached out by preheating to a judiciously selected temperature. In some cases use of PTTL glow peaks may be of advantage, since some of the peaks may not be produced in PTTL glow curve. This makes the peaks which appear in PTTL better resolved. For example in LiF:Mg, Ti the so-called peak 5 produced by PTTL is found to be free from peak 4, which in the γ irradiated samples superimposes on the lower temperature part of peak 5. In cases, in which more than one emitting species are found to be involved in spectral measurement of TL light, one may use the relevant monochromatically recorded glow peak. For example, in the blue-grain fluorite mineral, which is a popular dosimetry material due to its high sensitivity, one sees emission wavelengths of Ce, Tb, Dy, Er, and Sm in the glow peaks appearing in 75–500° C temperature range. Recording of monochromatic glow curves using selected wavelength of these emitters shows that the monochromatically recorded glow peaks are slightly displaced in temperature from the total light glow peak (Sunta [6]). Obviously one expects better accuracy in the measurements using monochromatic glow curves as compared to those using integral light.

(3) *Uncertainties in the methods based on peak shape and curve fitting techniques*—In the use of GO kinetics expression to characterize the glow peaks it is implicitly assumed that this expression encompasses all types of physically plausible systems which may apply to different glow peaks. This perception does not appear to be totally valid, since it is found that the quality of fit depends on the parameters of the system to which the GO expression is applied (Sunta et al. [24]). A "best fit" may not always be a proof of the accuracy of the results obtained. This may, for example, be obvious if the FOM value pertaining to the "best fit" is found to be high (FOM $>$ 1 %). Unfortunately many of the TL glow curve fitting exercises in the published literature do not present the value of FOM. The usefulness of the results of such fittings may be subject to doubt.

(4) *Effects of lacunae in GO kinetics expression*—Over and above the points discussed in the preceding paragraphs, it needs to be reminded here that the GO kinetics expression suffers from a fundamental lacuna (refer Sect. 3.3.4.1), according to which b and s' cannot have fixed values at all trap occupancies and at all T along the glow curve except when $b = 1$ or 2. In order to quantitatively determine the errors in the found values of E Sunta et al. [24] applied the GO kinetics model to the simulated glow peaks of three physical models namely the one trap—one recombination centre (OTOR) (refer Sect. 3.3.4), the noninteractive multi-trap system (NMTS) and the interactive multi-trap system (IMTS) models (refer Sect. 3.5). The input E value was kept same ($E = 1$ eV) in all the cases. The initial concentrations of the traps and the transition coefficients for retrapping and recombination and recapture in thermally disconnected deep traps (TDDTs) were changed in different sets of input parameters. Thus, glow peaks of different T_m and different shape factors were obtained. When these glow peaks were analyzed by GO peak fitting, the found values of E are seen to differ from the input value whenever the best fit value of b is found to be different from 1 and 2. The error in the found value of E is seen to increase as the fitted b value moves away from 1 and 2 (Sunta et al. [24]). Maximum value of error in the fitted E value is +8 % forthe fitting of the OTOR model glow curves and that for the glow curves of NMTS and IMTS models is −8 %. The errors are much greater for fitted b values <1 and >2. The maximum values of errors are seen at $b = 1.5$ for cases $1 < b < 2$. The b value 1.5 is farthest both from $b = 1$ and $b = 2$ when $1 < b < 2$. Interestingly the FOM value also is highest at $b = 1.5$ (Fig. 4.9). These results show that the quality of fit deteriorates as the best fit value of b moves away from 1 and 2. Errors as well as the FOM are very high when the b values are >2 and <1 indicating that such values of b are not meaningful. These findings once again show the drawback in the GO kinetics expression. However, in view of the numerous difficulties and uncertainties in various methods of glow curve analysis, as discussed in the preceding paragraphs, an error up to ±8 % in the found E value may not be considered too large. This may be a consolation for the glow curve analysis based on GO kinetics.

An interesting example that demonstrates the drawbacks of the fitting technique is shown in Fig. 4.10a. In this figure is shown the best GO fit of a simulated OTOR model glow peak whose input parameters are $E = 1$ eV, $s = 10^{12}$ s^{-1}, $N = 10^{12}$ cm^{-3}, $n_o = 10^{10}$ cm^{-3}, $R = 10^{-3}$ (where $R = A_n/A_h$) and $\beta = 1$ K s^{-1}. The analytical equation of the OTOR model (see (3.27)) is used to calculate this glow curve. The observed values of T_m and the peak symmetry factor μ and the best fitted values of E, b and the overall deviation of the fitted curve (dev. %) are given in the figure caption. Because of the high value of deviation (=3.751 %) found in fitting this glow peak, a two peak fit was tried on this glow curve. The result is shown in Fig. 4.10b. The fitted values of T_m, E, b, and dev. (%) are given in the figure caption for the two resolved peaks. It is interesting to see that the sum of the two best fitted peaks is much closer to the simulated single peak than the one peak

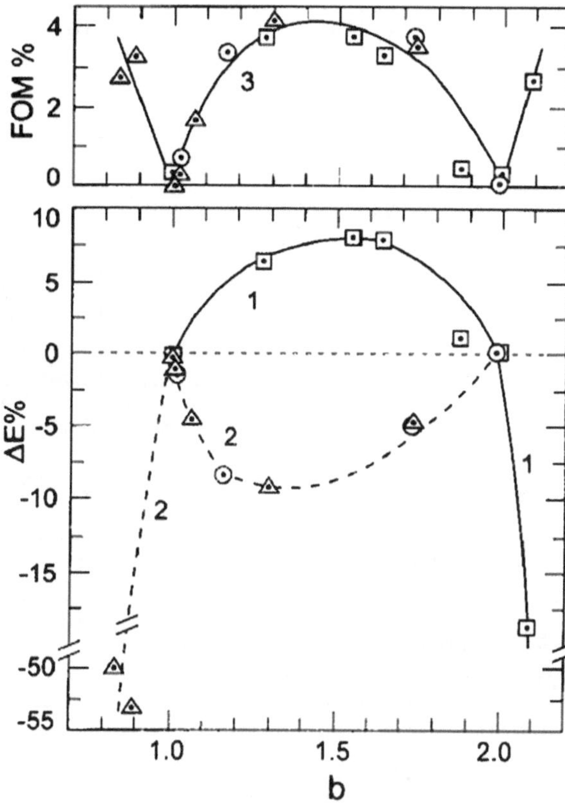

Fig. 4.9 Plots of ΔE versus b and FOM versus b, where ΔE is the error in the best fitted value of E when the simulated glow peaks of different physical models are fitted into GO kinetics expression. The different physical models for which glow curves are calculated are OTOR, NMTS and IMTS models (refer Chap. 3). The point's marked with ◇, Θ and Δ are respectively for these three models. The input values of parameter used in simulations are: $E = 1$ eV, $s = 10^{12}$ s^{-1}, $N = 10^{12}$ cm^{-3}. (after Sunta et al. [32, Fig. 1]). © IOP Publishing. Reproduced by permission. All rights reserved

best fit curve. This may be seen from the overall deviation from the actual data points. In the case of the two peak fit the deviation is only 0.255 % whereas the deviation in one peak GO best fit is 3.751 %. The stronger of the peaks in the two peak fitting has $b = 1$ and $E = 1.003$ eV. The second peak, which is about 1/3 in intensity of the first peak, fits best with $b = 2$ and $E = 1.755$ eV. This result obviously is an anomaly, since the simulated subject peak to which the fitting is applied has been calculated for an isolated single trap of the OTOR model. This anomaly takes place, because as explained earlier, the fitting results are always better with $b = 1$ and $b = 2$. This is because the GO expression is strictly valid only for $b = 1$ and $b = 2$, otherwise the true value of b varies continuously with T along the glow peak when the b value on the basis of shape of the peak differs from 1 and 2. As stated above the subject peak could be best fitted with $b = 1.227$, which resulted in E value higher by 6.4 % from its input value and an overall deviation of 3.751 %. This example shows that a "best fit" gives no guarantee of truthful results. It also highlights the importance of deciding the number of glow peaks before undertaking the fitting exercise on a multi-peak glow curve.

An example of a real sample which shows an increase in deviation or FOM as the best fitted value of b moves upward from 1 is seen in the work of Yossian and

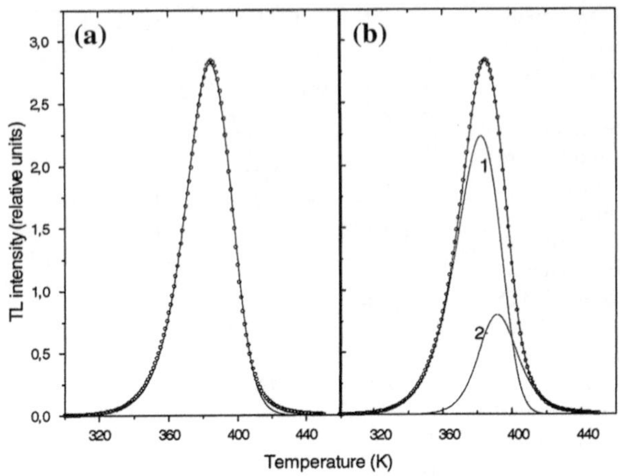

		T_m	E (eV)	b	Dev (%)
(a) One peak fit		384.9	1.064	1.277	3.751
(b) Two peak fit	**Peak 1**	382	1.003	1.106	0.255
	Peak 2	392	1.755	2.152	

Fig. 4.10 GO fitting of a simulated OTOR model glow peak. **a** One peak best fit **b** Two peak best fit. The glow peak is numerically calculated by using (3.27) (refer Sect. 3.3.4). The input parameter values are $E = 1$ eV, $s = 10^{12}$ s^{-1}, $N = 10^{12}$cm^{-3}, $R = 10^{-3}$ and $\beta = 1$ Ks^{-1}. The simulated glow peak has $T_m = 384.9$ K and $\mu = 0.432$. The best fitted Values of E, T_m and b in one peak and two peak best fits of this simulated peak together with the corresponding dev (%) are given above . (After Sunta et al. [33, Fig. 6]).

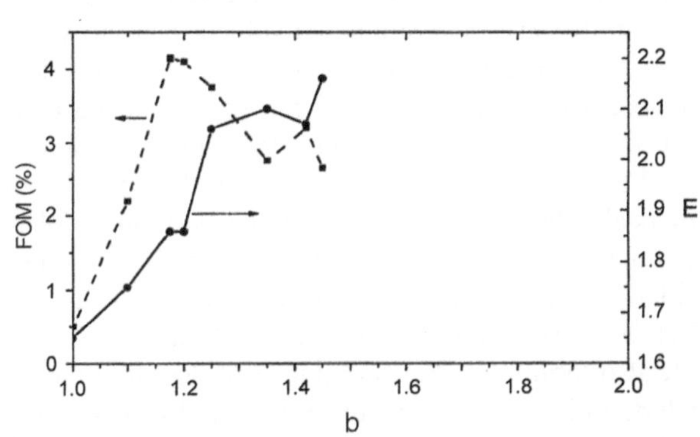

Fig. 4.11 Best fitted values of E and corresponding value of FOM both plotted against best fitted b in the GO fitting analyses of peak 5 of LiF:Mg, Ti (TLD-100) after post-irradiation annealing at 165 °C for various lengths of time. (after Sunta et al. [28, Fig. 6])

Horowitz [25, 26] who analyzed glow peak 5 of LiF:Mg, Ti (TLD-100) after post-irradiation annealing at 165 °C for various lengths of time. They found that the best fitted values of b as well as E increased with increase in annealing time. A close examination of their results shows that the FOM value also increased with increase in annealing time. An increase in annealing time is expected to result in decrease in trap occupancy. Therefore, a shift in T_m is expected with increase in annealing time when b is not equal to 1, but the authors found that the peak position remained unchanged. Further the mean life of the peak also did not show any change. No change in T_m with change in trap occupancy is a property which belongs exclusively to the FO kinetics. Yet these authors ignored this and concluded that the change in E value as a result of post-irradiation annealing of LiF:Mg, Ti (TLD-100) sample is unequivocal. Sunta et al. [28] have plotted the results of these authors which show that the so-called increase in E and b is associated with the deterioration in the quality of fit which is reflected by an increasing trend in FOM almost in parallel with the increase in the best fitted values of E and b. In Fig. 4.11 are plotted the results derived from Yossian and Horowitz [25]. To further investigate the question of any change in E by post-irradiation annealing, Sunta et al. [28] recorded the glow curve of the LiF:Mg, Ti (TLD-100) sample, irradiating, annealing, and reading the glow curves in the same way as done by Yossian and Horowitz. The glow curves were recorded for the samples with no post-irradiation anneal and for 5- and 20-min anneal at 165 °C. The glow peaks showed broadening and change in peak symmetry factor (δ/ω) with annealing time, but no change in T_m at a given heating rate. Sunta et al. also applied the different heating rates method to determine the E value for the unannealed and post-irradiation annealed samples and found that no change takes place in the E value by post-irradiation annealing. They then conjectured that the so-called peak 5 may not be a pure single energy peak and that the post-irradiation annealing produces minor modification in the defect distribution related with the peaks under the envelope of the so-called peak 5 without producing any change in T_m. The change in the shape of the assumed single peak after post-irradiation annealing, results in a change in b in the fitting exercise. It is well known that changes in defect distribution take place when samples are maintained at higher temperatures. In GO kinetics, parameters E and b are correlated. An increase in b is related to an increase in E when T_m is the same. It is therefore that the observed change in the shape without a change in T_m in the work of Yossian and Horowitz [25] resulted in a change in b together with a change in E.

4.9 Concluding Remarks

There has been a plethora of papers in the published literature in which glow curves are analyzed on the basis of the peak shape alone. Many workers seems to be either unaware or unmindful of the fact that the T_m of the non-first-order glow peaks increases with decease in trap occupancy (dose) and that a fixed T_m

independent of trap occupancy is a property belonging exclusively to the FO kinetics (see Fig. 4.1). These workers take the peak shape factor δ/ω as the sole criterion for deciding the kinetics order without paying attention to the peak temperature T_m. However, in the case of the annealed LiF:Mg, Ti (TLD-100) the authors Yossian and Horowitz [25, 26] knowingly ignored the observation that the T_m did not change after reduction of trap occupancy by post-irradiation annealing. This mindset of overdependence on peak shape for analyzing the TL glow curves, perhaps has a historical background. Chen [1] in his pioneering work to correlate the kinetics order value with the symmetry factor δ/ω of the glow peak, assumed $s'n_o^{b-1} = s$, a constant akin to the frequency factor with units s^{-1}. This clubbing of s', n_o , and b into a single constant altered the GO kinetics glow curve equation (refer (3.17) and (3.18) in Sect. (3.3.3), which rendered the temperature of the peak maximum T_m independent of n_o (dose) like in FO kinetics (refer 3.18) and also T_m became nearly independent of b. Later workers including those who wrote review articles and books on the theory of TL did not show the dependence of T_m on n_0 and b. The experimentalists therefore seem to have remained unaware or unmindful of the fact that if the T_m does not change with dose, the glow peak should belong to FO kinetics. If a glow peak belongs to any of the non-first-order (second order, GO, or mixed order) kinetics models, its T_m has to change with dose. Apparently, the paper by Sunta et al. [27] is the only publication which depicts dependence of T_m on n_0 and b in second order and GO kinetics. Later, the dependence of T_m on n_0 is shown by the same authors also for MO kinetics [28].

Coming back to the case of post-irradiation annealed LiF:Mg,Ti (TLD-100) it is interesting to see that later investigation by Weizman et al. [29] has shown that the so-called peak 5 of LiF:Mg,Ti (TLD-100) can be analyzed as consisting of three components, each of which is governed by FO kinetics. This example clearly shows that the weak hidden peaks under an apparently single peak may influence the main glow peak shape vitiating the analysis carried out on the basis of the shape factor. Secondly, the fact that this glow peak has been analyzed by assuming it as a single trap depth peak (the depth implicitly increasing with post-irradiation annealing) falling under general order kinetics and also as a combination of three subpeaks all falling under FO kinetics reflects an ever present uncertainty in the curve fitting method. An interesting case of more than one possible way of fitting a given glow curve is already given in Fig. 4.10 for a simulated glow peak. In this connection, a comment by Kelly et al. [30] may be relevant to mention here. They have expressed strong reservations about interpreting the experimental results uniquely if one starts from a position of complete ignorance of the parameters of the solid. Referring to their own simulated single trap depth broad glow peaks, they state that it can be analyzed also in terms of two or more trap depths in addition to the single trap depth used in the simulation. This comment is applicable also to the case of Fig. 4.10. The comment of Kelly et al., however, is made after looking at the simulated glow curves. In these simulations either the concentration of the recombination centers or their cross-sections or both may be so low that quasi-equilibrium (QE) approximation is not satisfied. It is under the non-QE conditions that such glow peaks are found. One cannot say that the situation could be as pessimistic as expressed by Kelly et al. in dealing with the real samples. In the case of simulated glow peaks, it is often seen that the parameter values (*N*, *M*, n_0, m_0, A_n, A_m, A_h, *s*, etc.) are arbitrarily chosen without regard to whether the chosen values would be applicable to the real materials. A striking example of such a case is seen in the work of Lewandowki and McKeever [31] which gives numerically computed glow curves using the total trap concentration values equal to 10^9 m^{-3}. The resulting TL glow profiles do not have a shape worthy of being named as glow curves. Such low values of defect concentrations are unimaginable in real materials even if one takes purest possible materials and as much vibration free equipment as possible to grow the crystals. Any generalized conclusions drawn from such simulations may only mislead the readers (Sunta et al. [32]). It is imperative that the parameter values used in the simulations should be physically realistic remembering the fact that the computed results, which include the shape of the glow peaks, depend on the input parameter values (refer also to Sects. 5.3 and 5.4 in Chap. 5). The importance of choosing realistic parameter values in simulations and the plausible range of the parameters in real materials is discussed later in Chap. 5 (refer Sect. 5.5).

If we compare different methods of determining the *E* value it appears that the initial rise method may score over other methods provided one is able to catch the initial rise part free from any contribution from nearby glow peaks in multi-peak glow curves (Sunta et al. [33]). This method is applicable irrespective of the kinetic order. It does not involve any approximation like in peak shape and peak fitting methods. Even in the case of multi-peak and composite glow curves, recording of multiple initial rise curves leading to a ladder type *E* versus *T* graph may produce reasonable values of *E* (see Sect. 4.3.1). Further, it may be advisable for an experimentalist to use more than one method to analyze the glow curves of the given sample. A comparison of results from different methods may give an educative experience.

References

1. R. Chen, J. ElectrochemSoc. Solid State Sci. **116**, 1254 (1969)
2. N. Takeuchi, K. Inabe, H. Nanto, Solid State Commun. **17**, 1267–1269 (1975)
3. F. Urbach, Wien. Ber. II a **139**, 363 (1930)
4. G.F.J. Garlick, A.F. Gibson, Proc. Phys. Soc. London **60**, 574 (1948)
5. N. S. Rawat, M. S. Kulkarni, D. R. Mishra, B. C. Bhatt, C. M. Sunta, S. K. Gupta, D. N. Sharma, Nucl. Instrum. Methods Phys. Res. B **267**, 3475(2009)
6. C.M. Sunta, Radiat. Prot. Dosimetry. **8**, 25 (1984)
7. H. Gobrecht, D. Hoffman, J. Phys. Chem. Solids **27**, 509 (1966)
8. J. Nahun, A. Halperin, J. Phys. Chem. Solids **24**, 823 (1963)
9. Y. Kirsh, S. Shoval, P.D. Townsend, Phys. Stat. Sol. A **101**, 253 (1987)
10. J.K. Srivastava, S.J. Supe, J. Phys. D: Apppl. Phys. **16**, 1813 (1983)
11. P. Braunlich, J. Appl. Phys. **38**, 2516 (1967)
12. C.M. Sunta, W.E.F. Ayta, T.M. Piters, S. Watanabe, Radiat. Meas. **30**, 197 (1999)

13. C.M Sunta, V.N. Bapat, in *Specialist Seminar on Thermoluminescence Dating OXFORD (1980)* PACT vol 6 (1982), p. 252
14. W. Hoogenstraaten, Philips Res. Rep. **13**, 515 (1958)
15. T.S.C. Singh, P.S. Mazumdar, R.K. Gartia, J. Phys. D Appl. Phys. **23**, 562 (1990)
16. R. Chen, S.A.A. Winer, J. Appl. Phys. **41**, 5227 (1970)
17. J.T. Randall, M.H.F. Wilkins, Proc. R. Soc. London Ser A, **185**, 365 (1945)
18. C.E. May, J.A. Partidge, J. Chem. Phys. **40**, 1401 (1964)
19. R. Chen, S. W. S. McKeever, *Theory of Thermoluminescence and Related Phenomena* (Singapore, World Scientific, 1997), pp. 110–121
20. A. Halperin, A.A. Braner, Phys. Rev. **117**, 408 (1960)
21. S.J. Singh, R.K. Gartia, P.S. Mazumdar, J. Phys. D Appl. Phys. **22**, 467 (1989)
22. D. Shenker, R. Chen, J. Comput. Phys. **10**, 272 (1972)
23. D.W. Zimmermann, Archaeometry **12**, 29 (1970) (quoted by M.J. Aitken Physics Reports, Archaeol. Involvements Phys. **40**C(5), 285 (1978)
24. C.M. Sunta, W.E.F. Ayta, J.F.D. Chubaci, S. Watanabe, Radiat. Meas. **35**, 47 (2002)
25. D. Yossian, Y.S. Horowitz, J. Phys. D Appl. Phys. **28**, 1495 (1995)
26. D. Yossian, Y.S. Horowitz, Radiat. Meas. **27**, 465 (1997)
27. C.M. Sunta, W.E.F. Ayta, R.N. Kulkarni, T.M. Piters, R. Chen, S. Watanabe, Radiat. Prot. Dosim. **71**, 93 (1997)
28. C.M. Sunta, W.E.F. Ayta, J.F.D. Chubaci, S. Watanabe, J. Phys. D Appl. Phys. **38**, 95 (2005)
29. Y. Weizman, Y.S. Horowitz, L. Oster, J. Phys. D Appl. Phys. **32**, 2118 (1999)
30. P. Kelly, M.J. Laubitz, P. Braunlich, Phy. Rev. **B4**, 1960 (1971)
31. A.C. Lewandowki, S.W.S. McKeever, Phys. Rev. B **43**, 8163 (1991)
32. C. M. Sunta, W. E. F. Ayta, T. M. Piters, R. N. Kulkarni, S. Watanabe, J. Phys. D Appl. Phys. **32**, 1271–1275 (1999)
33. C.M. Sunta, W.E.F. Ayta, R.N. Kulkarni, T.M. Piters, S. Watanabe, J. Phys. D Appl. Phys. **30**, 1234–1242 (1997)

Chapter 5
The Quasi-Equilibrium Problem in Thermoluminescence

Abstract The chapter begins with defining the quasi-equilibrium (QE) in thermoluminescence (TL). It is shown that it is the values of the system parameters namely the concentration N of the traps and the recombination centers (RC) and their cross-sections σ which solely determine whether or not the QE condition would be satisfied. This dependence may be used to answer the question whether QE is satisfied in the TL of real materials. Greater are the values of these parameters better it is for the QE condition. A function of N and σ is derived which defines the lower limit of these parameters for satisfying the QE condition. Some earlier workers had concluded that the QE may be satisfied only when recombination rate dominates over the retrapping rate during the glow curve recording. In this chapter this condition is shown to be unimportant. The importance of the recombination lifetime τ_{rec} is highlighted. Smaller is the value of τ_{rec} better it is for the QE to be achieved. It is found that the condition $\tau_{rec}/T_R < 10^{-3}$ sets the upper limit of τ_{rec} in any glow peak, where T_R is the time span of the glow peak being measured. It is found that QE depends also on the heating rate β. When β is increased above a certain limit, the glow peak may begin to widen. At this point the glow peak enters from QE into non-QE domain. It is suggested that this property may be used to test whether a glow peak is being recorded under the QE condition. The limiting value of β above which this effect may be seen depends on the parameters of system under investigation. As an illustration, this test is applied to the glow peak No. 5 of LiF-TLD phosphor.

5.1 Definition

'Quasi' is a Latin word. It means very nearly or almost. In the analytical models of thermoluminescence (TL) namely the first-order (FO) kinetics model of Randall and Wilkins (RW) [1], second-order (SO) kinetics models of Garlick and Gibson (GG) [2], general-order (GO) kinetics model of May and Partrige (MP) [3], and

C. M. Sunta, *Unraveling Thermoluminescence*, Springer Series in Materials Science 202, DOI: 10.1007/978-81-322-1940-8_5, mixed-order (MO) kinetics model of Visocekas [4], the recombination plus retrapping rate during heating are assumed to be equal to the thermal stimulation rate which means that the excited charge carriers during heating to read the TL are assumed to instantly relax either into the available recombination centers (RC) or into the vacant traps. This means that the equilibrium of the excited state (conduction band, assuming that the traps are of electron type) existing before the start of the heating the sample to read its TL remains undisturbed. Mathematically this is expressed as $dn_c/dt = 0$, where n_c is the concentration of the electrons in the conduction band. In reality, however, the charge carriers have a finite life time in the excited state. The excited carriers therefore tend to stay in that state for a time length which depends on the life time of the said state. Thus there may take place some build up of the carriers in the excited state. However if this build up is negligible in comparison to the population n of the active traps at all temperatures T during heating the equilibrium of the excited state is assumed to be undisturbed for all practical purposes. This is called the quasi-equilibrium (QE) approximation. Mathematically the QE approximation for the simple one trap-one recombination centre (OTOR) model is expressed as $n_c \ll n$ and $dn_c/dt \ll dn/dt$ or $dn_c/dt \approx 0$. Analytical expressions are the backbone of the TL theory. Various authors [5–10] have therefore attempted to fix the criteria that is, the extent to which the disequilibrium may be allowed so that the analytical expressions may be assumed valid for all practical purposes. Essentially this means to determine how small the smaller of the quantity should be in comparison to its greater counterpart in the above inequalities. To understand the QE approximations we need to describe the transport of the released charge carriers from the traps to the RC during the heating in terms of the rate equations in an exact manner for chosen physical models. QE approximations may then be applied to the rate equations to arrive at the analytical expressions. In Chap. 3 while discussing the various physical models of TL using the rate equations, the QE approximations are amply used. The QE approximations may be summed up as following for the three physical models, namely the OTOR model (refer Sect. 3.3.4), the noninteractive multitrap system (NMTS) model (refer Sect. 3.5.3) and the interactive multitrap system (IMTS) model (refer Sect. 3.5.2).

(1) OTOR model, $n_c \ll n$ and $|dn_c/dt| \ll |dn/dt|$
(2) NMTS model, $n_c \ll (n + M)$ and $|dn_c/dt| \ll |dn/dt|$
(3) IMTS model, $n_c \ll (n + m)$ and $|dn_c/dt| \ll |-dn/dt - dm/dt|$

The meaning of various terms in these inequalities is already given in Chap. 3 and also in the Appendix 1. The above so-called QE approximations are only qualitative in nature. To be of practical value one should know how much smaller the quantities should be in comparison to their respective greater counterparts so that it could be neglected. This question has been addressed by Sunta et al. [9]. According to these authors, to define QE quantitatively one should fix a limiting value of the ratio between the respective smaller and larger quantities in the above inequalities. This limiting value logically should be related to the accuracy achievable in the TL intensity measurements. Experience has it that the accuracy in the experimental measurements of TL may not be better than 1 %. One may therefore say that QE condition may be considered to be well satisfied if the ratio between smaller and the greater quantities in the above inequalities is $\leq 10^{-2}$ at all T during the heating phase. For example, n_c/n and $|dn_c/dt|/|dn/dt|$ may be $\leq 10^{-2}$ in OTOR model. Using the rate equations, Shenker and Chen [5] calculated the TL and TSC glow curve for selected set of input parameters without applying the QE approximations. They concluded that the QE approximation is valid only when the ratio $|dn_c/dt|/|dn/dt| \leq 10^{-3}$ throughout the active temperature range of the glow peak. Often at the high temperature end of the glow peak this ratio does not hold. For such cases, Chen and Pagonis [11] suggest that instead of comparing dn/dt with dn_c/dt, one may better compare the rate of adding electrons to the conduction band to the depletion of electrons from the said band by retrapping and recombination. Kelly et al. [6] examined the validity of QE in greater detail by using various parametric combinations in a model involving a single trap depth in the presence of the TDDT and comparing the exact numerical results with those obtained analytically after applying QE conditions. They concluded that the validity of the QE assumption depends critically on the total concentration N of active traps. In particular, they showed that the QE assumption is adequate if $N > 10^{15}\text{cm}^{-3}$. Sunta et al. [8] have later shown that the QE depends not only on N but also on the capture cross-section of the traps. This is discussed in greater detail later in this Chapter.

5.2 An Alternative Definition for QE and Its Meaning with Reference to Its Conventional Definition

Conventionally n_c/h and $|dn_c/dt|/|dn/dt|$ ratio values are the direct measures of the level of QE in a system, where h = number of RC which is equal to n in OTOR, equal to $(n + M)$ in NMTS and equal to $(n + m)$ in IMTS. Lewandowski and Mckeever (LM) [7] introduced a new approach. They introduced two new functions called $Q(T)$ and $P(T)$ to account for QE and kinetic order (KO) respectively. As will be seen in the following the $Q(T)$ function accounts only for $|dn_c/dt|/|dn/dt|$ ratio. The computations of Lewandowski and Mckeever [7] are based on the NMTS model in which they assumed $M \gg n$, which renders $(n_c/h) \ll 1$ at all T, where $h = (n + M)$. Thus one of the conditions in the QE approximations is already met by the chosen parameters in this system. In any other physical model, the condition $(n_c/h) \ll 1$ may need to be verified. LM defined $Q(T)$ and $P(T)$ functions in terms of three reaction rates during heating phase namely excitation rate (R_{ex}), retrapping rate (R_{ret}) and recombination rate (R_{rec}) viz,

$$Q(T) = \frac{(R_{ex} - R_{ret})}{R_{rec}} \tag{5.1}$$

$$P(T) = R_{ret}/R_{rec} \tag{5.2}$$

where

$$R_{ex} = ns\exp(-E/KT) \tag{5.3}$$

$$R_{ret} = n_c(N-n)A_n \tag{5.4}$$

$$R_{rec} = n_c(n+n_c)A_h \text{ in OTOR} \tag{5.5a}$$

$$R_{rec} = n_c(n+n_c+M)A_h \text{ in NMTS} \tag{5.5b}$$

$$R_{rec} = n_c(n+n_c+m)A_h \text{ in IMTS.} \tag{5.5c}$$

The values of *n, m,* and n_c change with temperature during the heating. The three functions R_{ex}, R_{ret}, and R_{rec} are therefore a function of temperature *T* and as a result their values change with *T* along the span of the glow curve. Consequently, the *Q*(*T*) and the *P*(*T*) also are dependent on *T*. It is to be noted here that *P*(*T*) is intended to replace the empirical KO parameter *b*. The latter parameter is assumed as a constant in the GO kinetics model. Seen in the context of *P*(*T*) it implies that *b* is a kind of average in place of the variable parameter *P*(*T*). The effect of the nonconstancy of the parameter *b* has been discussed earlier in Sect. 3.3.4.1. However, the advantage with the parameter *b* is that it can be directly measured from the shape of an experimental glow curve and then used for its analysis, not withstanding its limitation. In contrast, there is no way to know the *P*(*T*) values for a given experimental glow curve, since *P*(*T*) is dependent on parameter values *n, m, N, M,* A_n, A_m, and A_h which are unknown for an experimental glow curve. The *P*(*T*) function thus has no practical applicability. It is useful only in studying the behavior of the simulated TL glow curves. The simulations, however, are useful, since they give insight into many aspects of the TL phenomenon. The function *Q*(*T*) too has its own limitation. A comparison of the *Q*(*T*) values with the conventional QE function, $|dn_c/dt|/|dn/dt|$ ratio at various heating rates shows that *Q*(*T*) is not sensitive enough as compared to the $|dn_c/dt|/|dn/dt|$ ratio. It is shown later that when the ratio $|dn_c/dt|/|dn/dt|$ changes by a decade and the system enters from the QE into a non-QE domain due to the change in heating rate, the *Q*(*T*) value shows no perceptible change (see Fig. 5.8).

5.2.1 Reconciliation of *Q*(*T*) with the Conventional Definition of QE

Since R_{rec} means the TL intensity *I* and (R_{ex}–R_{ret}) means –d*n*/d*t* (see (3.22), we may write (5.1) as

$$Q(T)I = -\mathrm{d}n/\mathrm{d}t. \tag{5.6}$$

According to (3.24), we have

$$I = -\mathrm{d}n/\mathrm{d}t - \mathrm{d}n_c/\mathrm{d}t.$$

We see that when $Q(T) = 1$ in (5.6) and $|\mathrm{d}n_c/\mathrm{d}t| << |\mathrm{d}n/\mathrm{d}t|$ or $\mathrm{d}n_c/\mathrm{d}t \approx 0$ in (3.24) we get $I = -\mathrm{d}n/\mathrm{d}t$. Thus we see that $Q(T) = 1$ has the same effect on the expression for TL intensity I as that of $\mathrm{d}n_c/\mathrm{d}t \approx 0$. We may thus say that $Q(T) = 1$ is an alternative condition to describe the QE approximation.

5.2.2 Inferences from the Use of Q(T) and P(T)

From (5.1) and (5.2) it is seen that

$$Q(T) + P(T) = R_{ex}/R_{rec} \tag{5.7}$$

and

$$Q(T)/P(T) + 1 = R_{ex}/R_{ret}. \tag{5.8}$$

In (5.7) and (5.8) Lewandowski and Mckeever [7] consider two types of cases: (i) a fast retrapping case with $Q(T) = 1$ and $P(T) \gg 1$. According to (5.8), it implies that under this case $R_{ex} = R_{ret}$. If this were strictly true, the active traps would not empty since in such a case the charge carriers would simply recycle between the traps and their excited states during heating. Furthermore if R_{ex} is equal to R_{ret}, we get $Q(T) = 0$ in (5.1), contrary to the initial premise $Q(T) = 1$. Thus TL peaks would not be observed. Therefore, according to these authors, for the peaks to be seen the conditions $P(T) \gg 1$ and $Q(T) = 1$ together cannot both be true. In the case (ii) the authors consider a slow retrapping case, i.e., $P(T) \ll 1$ together with $Q(T) = 1$. Under this condition, one would, according to (5.7), have $R_{rec} = R_{ex}$, which is exactly the condition in RW model (see Sect. 3.3.1). Thus according to the logic presented here, the TL peaks can be observed only under the condition $P(T) \ll 1$ and $Q(T) = 1$. These arguments appear quite logical, but it must be remembered that they are only qualitative in nature. It will be seen in the following that when we consider the (5.1) and (5.2) quantitatively using the actual values of various parameters (N, M, A_n, A_h, etc.) in (5.3)–(5.5a–c), the results may turn out to be quite different. Nevertheless Lewandowski and Mckeever [7] have supported their above stated conclusions regarding the incompatibility of $P(T) \gg 1$ and $Q(T) = 1$, with two examples one having $P(T) \ll 1$ and other $P(T) \gg 1$. These examples named by the authors [7] as set A and set F use the following parameter sets to compute numerically the TL glow curves

Set A: $E = 0.3$ eV, $N = 10^9$ m^{-3}, $M = 10^{12}$ m^{-3}, $S(T) = 1.6 \times 10^{-17} T^{-3/2}$ m^2, $R(T) = 10^{-15} T^{-3/2}$ m^2.

Set F: $E = 0.3$ eV, $N = 10^9$ m^{-3}, $M = 10^9$ m^{-3}, $S(T) = 1.6 \times 10^{-17} T^{-3/2}$ m^2, $R(T) = 10^{-19} T^{-3/2}$ m^2.

where M = concentration of the thermally disconnected deep traps in the NMTS model (see Sect. 3.5.3) used by these authors. In this model, $h(T) = (M + n(T))$. The parameters $S(T)$ and $R(T)$ are capture cross-sections of the active traps and the RC, respectively. In this book we designate these parameters by σ_n and σ_h, respectively. The authors [7] have accounted for their temperature dependence as shown by $T^{-3/2}$ factor in each set. The temperature dependence though could have been ignored, since it does not make a major difference in the glow curves. The escape frequency factor s is taken to be equal to $N_c S(T) v(T)$ (refer Sect. 3.2.1), where $N_c(T)$ is the effective density of states (approx. 10^{25} m^{-3}) in the conduction band and $v(T)$ is the free electron thermal velocity (approx. 10^5 ms^{-1}). TL glow curves were calculated by these authors for various n_o/N values along with their corresponding $Q(T)$ and $P(T)$ profiles for set A as well as for set F. Their results show that for Set A in which $P(T) \ll 1$, the glow curves are properly shaped nearly like those of the FO kinetics (refer Fig. 2 in [7]). However, QE is not fully satisfied. The value of $Q(T)$ ranges from about 1.9–1 in the first half of the glow curve and from 1 to 0 in the second half. In contrast to set A, in Set F in which $P(T) \gg 1$ (refer Fig. 4 in [7]), the glow curves do not even have a peak like shape. These results are taken by Lewandowski and Mckeever [7] as the proof of the above stated conclusion that $Q(T) = 1$ and $P(T) \gg 1$ together do not form a self-consistent set of conditions. The authors further emphasized that the traditional fast retrapping ($P(T) \gg 1$) TL equations which include those of SO, GO, and MO kinetics (see Chap. 3) all of which use the QE approximation ($Q(T) = 1$), should be considered worthless. This indeed is a sweeping conclusion. Sunta et al. [8] have later shown that this conclusion is erroneous and that QE condition may be achieved irrespective of whether the process involved during heating is a slow retrapping type($P(T) \ll 1$) or a fast retrapping($P(T) \gg 1$) type. Slow and fast are relative terms. What is important is the absolute value of the recombination rate even if it is slower than the retrapping rate. Let us take an example from the GG model in which $R_{rec} \ll R_{ret}$. Assume $N = 10^{20}$ cm^{-3}, $n_0 = 10^{18}$ cm^{-3}, $A_n = A_h = 10^{-12}$ cm^{3}s^{-1} (where $A_n = \sigma_n v$ and $A_h = \sigma_h v$), $E =$ 1 eV, $s = 10^{12}$ s^{-1} and $\beta = 1$ Ks^{-1}. The glow curve and the $Q(T)$ and $P(T)$ are calculated numerically using the rate equations of the OTOR model (see 3.22–3.24). The results are shown in Fig. 5.1. These show $Q(T) = 1$ and $P(T)$ values varying from 10^2 to 10^4. The value of $P(T)$ at $T = T_0$ where $n = n_0 = 10^{18}$ and $N = 10^{20}$ may be easily seen to be $P(T) = (N-n)/n = 10^2$ without going through the involved numerical computations. The $P(T)$ value increases progressively as n decreases with increasing T in the second half of the glow peak. This result is clearly contrary to the conclusion of Lewandowski and Mckeever [7] that $P(T) \gg 1$ and $Q(T) = 1$ together cannot be true for a glow peak to be seen. It may be seen that the glow peak is emitted in a time span of about 200 s at the heating

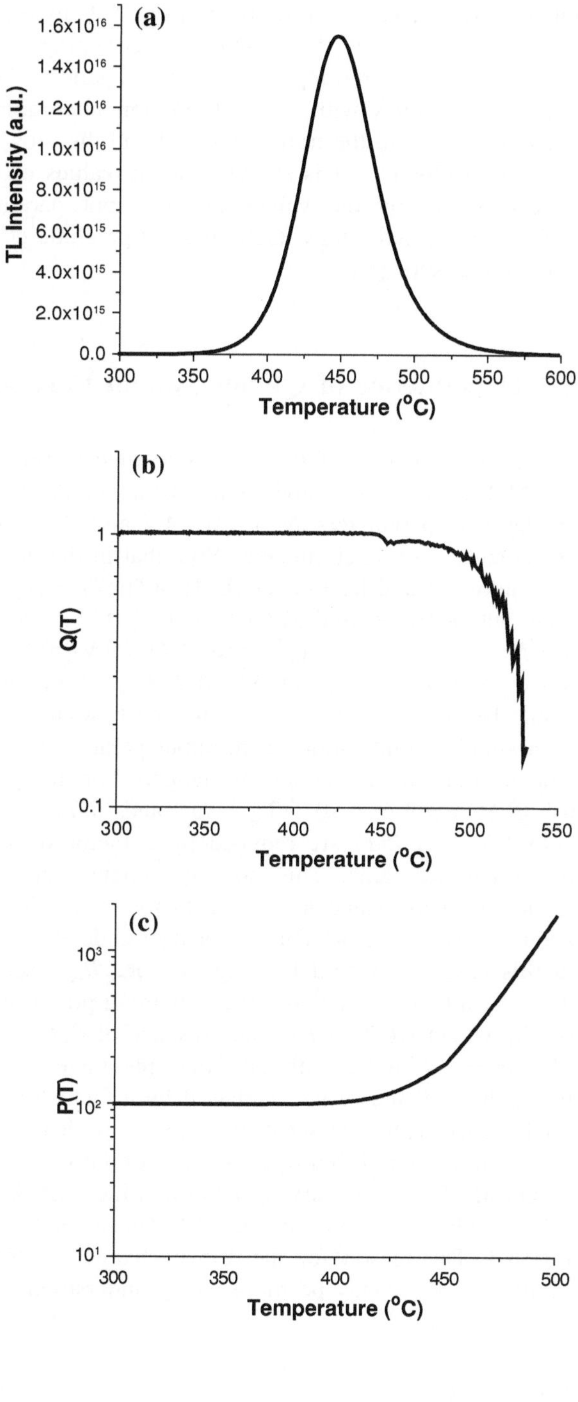

Fig. 5.1 **a** Profile of a GG model glow curve: input parameters are: $E = 1$ eV, $s = 10^{12}\ s^{-1}$, $N = 10^{20}\ cm^{-3}$, $n_0 = 10^{18}\ cm^{-3}$, $A_n = A_h = 10^{-12}\ cm^3\ s^{-1}$, heating rate $= 1\ K\ s^{-1}$. **b** Profile of Q(T) for the glow peak. **c** Profile of P(T) for the glow peak

rate of 1 K s^{-1}(see the glow curve in Fig. 5.1) by recombination transitions whose rate is 10^2–10^4 times slower than the retrapping rate. Over 99 % of the excited carriers which get retrapped are excited again as the cycle of excitation and re-trapping continues with a small fraction undergoing recombination after each excitation step, till the trap is emptied and the glow curve emitted fully.

In the following, it is shown that the values of the $Q(T)$ and $P(T)$ functions depend solely on the values of the input parameters of the given system. A $P(T) \gg 1$ condition by itself does not preclude $Q(T) = 1$, nor does a $P(T) \ll 1$ condition ensure $Q(T) = 1$.

5.3 Dependence of QE on System Parameters

As far as the values of $P(T)$ are concerned in set A and set F of Lewandowski and Mckeever [7] it is obvious that these values depend on the numerical values of the five parameters N, n, M, A_n, and A_h. Further when $n \ll N$ and M, dependence on n is eliminated. Note that in the parameter nomenclature used by Lewandowski and McKeever (LM) $S(T)v(T) = A_n$ and $R(T)v(T) = A_h$. In set A, these authors have used $M \gg N$ and $A_h \gg A_n$. On the other hand in Set F, they use $M = N$ and $A_h \ll A_n$. Since $P(T) = (N-n)A_n/(M + n)A_h$, one would obviously get $P(T) \ll 1$ in set A and $P(T) \gg 1$ in set F. In the present context it would be an interesting exercise to find how the values of $Q(T)$ function change by a simple modification in the input parameter values of set A and set F such that the $P(T)$ values remain unchanged. For this purpose we modified set A by using $S(T) = 1.6 \times 10^{-21}T^{-3/2}$ m^2 and $R(T) = 10^{-19}T^{-3/2}$ m^2, which means both $S(T)$ and $R(T)$ are reduced by a factor of 10^4 with all other parameters remaining the same. This does not change the value of $P(T)$. However the reduction in the value $R(T)$ by a factor of 10^4 leads to an increase in $Q(T)$ by about the same factor. This is because under the condition $P(T) \ll 1$ prevailing in this case, we would have $Q(T) = R_{ex}/R_{rec}$ (see 5.1). The R_{rec} having gone down by a factor of 10^4 and R_{ex} remaining practically unaffected, in the modified set A, we get $Q(T) \gg 1$ in the presence of $P(T) \ll 1$. In modified set F, we use $M = N = 10^{16}$ m^{-3}, with all other parameter values remaining same, which means both N and M are increased by a factor of 10^7 in comparison to original set F. Once again we see that there is no change in the value of $P(T)$ but the $Q(T)$ values are reduced by about the similar order of magnitude as the change in N and M. The glow curves of the modified set A together with their $Q(T)$ and $P(T)$ profiles are shown in Fig. 5.2. Those for the modified set F are shown in Fig. 5.3. The reason for the above changes in the glow curves and the $Q(T)$ profiles of set F may be more easily understood if we write $Q(T)$ as

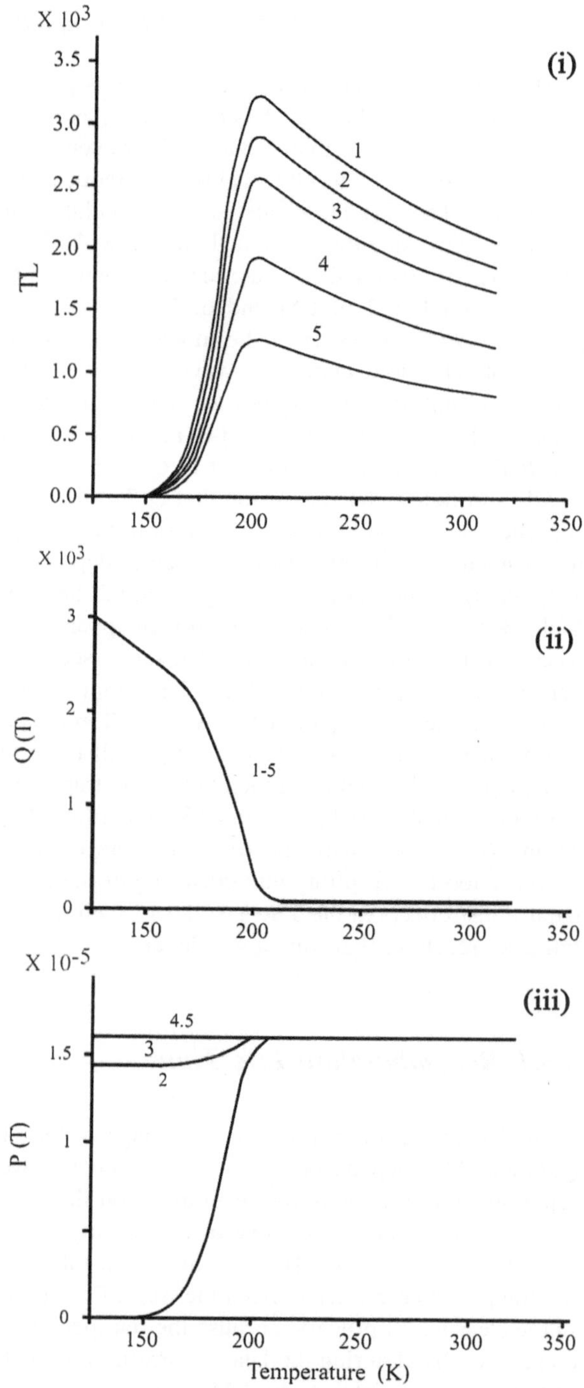

Fig. 5.2 **i** TL intensities as a function of temperature (K) for the modified set A (decreasing ($S(T)$ and $R(T)$ by a factor of 10^4) of Lewandowsky and McKeever [7]. Input parameters are $E = 0.3$ eV, $N = 10^9$ m^{-3}, $M = 10^{12}$ m^{-3}, $S(T) = 1.6 \times 10^{-21} T^{-3/2}$ m^2, $R(T) = 10^{-19} T^{-3/2}$, heating rate = 4 K min^{-1}. Glow curves 1 to 5 are for $n_0/N = 1, 0.5, 0.1, 0.05$ and 0.01 respectively. Intensities of curves 2, 3, 4 and 5 are after multiplication of actual values by factors 9, 80, 600, and 4000 respectively for plotting. **ii** $Q(T)$ values as a function of temperature for the glow curves. **iii** $P(T)$ values as a function of temperature for the glow curves

$$Q(T) = R_{ex}/R_{rec} - R_{ret}/R_{rec}.$$

The value of second term which represents $P(T)$ is same at its maximum value in modified set F (when one has $n \ll N, M$) as in original set F. In the first term, R_{ex} does not undergo a major change. However the value of R_{rec} having increased by a factor of 10^7, the $Q(T)$ value gets reduced by about the similar order of magnitude. The TL glow curves are calculated numerically for the modified parameter sets using the same method as used by Lewandowski and Mckeever [7]. The values of $Q(T)$ and $P(T)$ are also calculated in the same way. Comparison of Fig. 5.2 with Fig. 2 of LM and of Fig. 5.3 with Fig. 4 of LM shows that the profiles of $P(T)$ are same in the modified sets as in original sets A and F. The nature of TL glow curves, however, is reversed. The glow curves of set A (modified) and set F (modified) are similar to those of original set F and set A, respectively. It is noteworthy that in modified set A, $Q(T) \gg 1$ in spite of the fact that $P(T) \ll 1$ and in modified set F, $Q(T) = 1$ even when $P(T) \gg 1$. This result clearly shows that the conclusion that $Q(T) = 1$ and $P(T) \gg 1$ do not form a self-consistent set of conditions, is not valid. In fact the $Q(T)$ and the $P(T)$ functions are independent of each other and their values depend on the basic parameters (N, M, $S(T)$, $R(T)$) of the system. The logic behind the conclusion of Lewandowski and Mckeever [7] as discussed above, though apparently reasonable, fails because it is based on the ratios of three different rates (see (5.3)–(5.5a–c)) each of which is dependent on the absolute values of the input parameters. The authors [7] used very low values of the parameters which fall in the unphysical range, rendering the recombination rates too low, with the result that under certain sets of such low-value parameters such as in their set F, the traps could not empty within the time duration normally used in TL and TSC recording. In particular, when the values of M and N are too small, the transition rate of the excited carriers into the RC becomes too low resulting into crowding of the excited carriers in the delocalized band. This reflects in the TSC profiles of their original set F which do not show a fall after reaching their maximum level.

5.3.1 Recombination Life Time

To understand more clearly how the shape of the glow curves, and the values of $Q(T)$ and $P(T)$ depend on the system parameters we may describe the trap emptying process in terms of the recombination lifetime τ_{rec}, of the charge carriers in the conduction band. The value of τ_{rec} is simply the inverse of the recombination rate of the free carriers. The recombination rate gives the rate of the clearing off the trapped charge carriers from the traps. Retrapping transitions are not included in the clearing off process, because the retrapped charges are thrown back into the conduction band during the heating. Presuming the QE condition the value of τ_{rec} in the NMTS model used by LM is given by

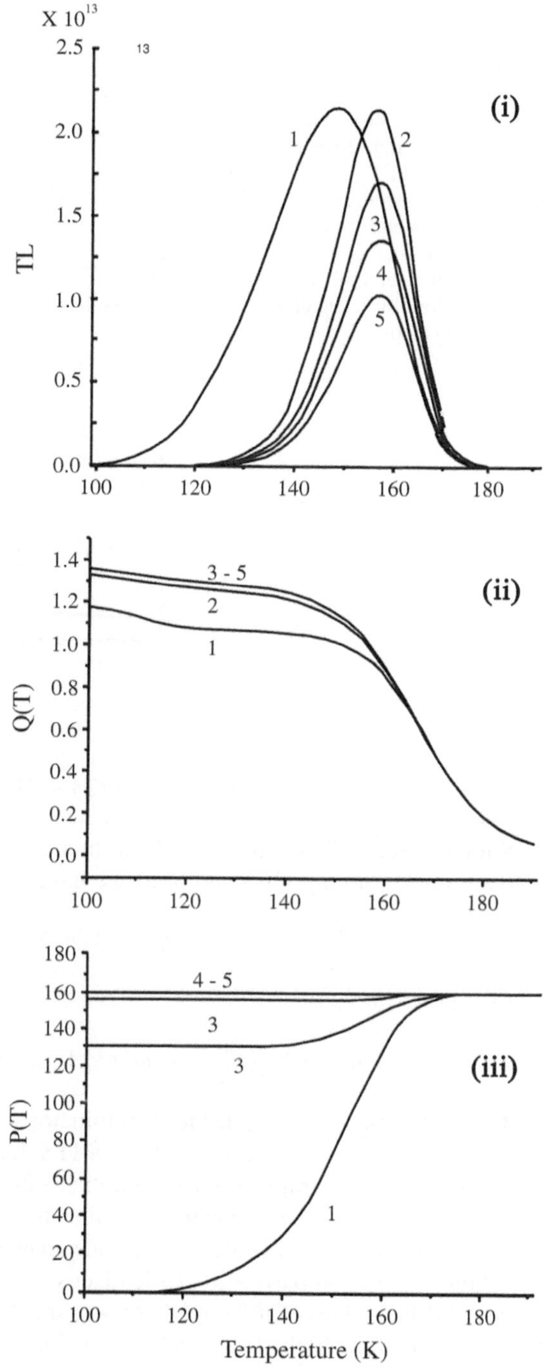

Fig. 5.3 **i** TL intensities as a function of temperature (K) for the modified set F (increasing the values of N and M by a factor of 10^7) of Lewandowsky and McKeever [7]. Input parameter values are: E= 0.3 eV, $N = M = 10^{16}\text{m}^{-3}$, $S(T) = 1.6 \times 10^{-17}T^{-3/2}$ m^2, $R(T) = 10^{-15}T^{-3/2}$ m^2, heating rate = 4 K min^{-1}. Glow curves 1 to 5 are for n_0/N = 1, 0.5, 0.1, 0.05 and 0.01 respectively. Intensities of curves 2, 3, 4 and 5 are after multiplication of actual values by factors 6, 50, 400 and 3000 respectively for plotting. **ii** $Q(T)$ values as a function of temperature for the glow curves. **iii** $P(T)$ values as a function of temperature for the glow curves

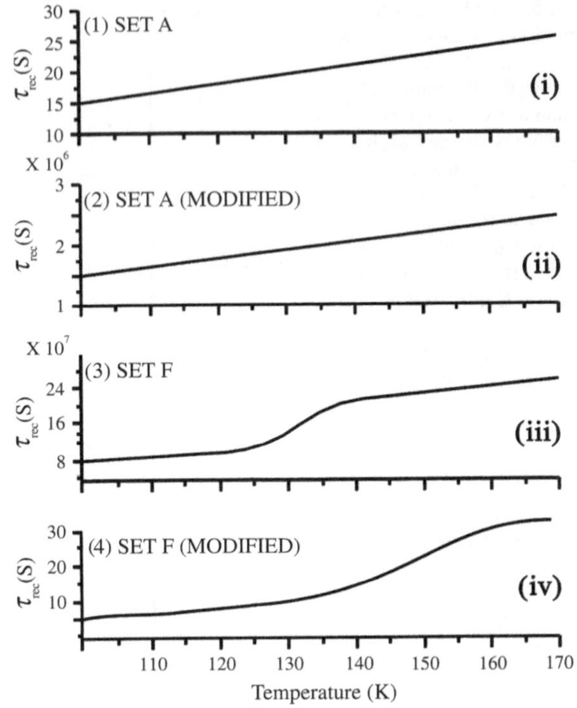

Fig. 5.4 **i** τ_{rec} as a function of temperature (K) during the glow curve reading of set A [7]. **ii** τ_{rec} as a function of temperature (K) during the glow curve reading of set A [7] (modified). **iii** τ_{rec} as a function of temperature (K) during the glow curve reading of set F [7]. **iv** τ_{rec} as a function of temperature (K) during the glow curve reading of set F [7] (modified)

$$\tau_{rec} = 1/[(n(T) + M)A_h]. \tag{5.9a}$$

Similarly for OTOR model in which the number of RC is equal to that of filled active traps (assuming QE condition) we have

$$\tau_{rec} = 1/n\,(T) \tag{5.9b}$$

and for IMTS model

$$\tau_{rec} = 1/[\{n(T) + m(T)\}A_h + \{M - m(T)\}A_m]. \tag{5.9c}$$

The term $\{M{-}m(T)\}A_m$ in the denominator represents the nonradiative relaxation probability. (Refer Sect. 3.5.2 for IMTS and NMTS models).

In order that the traps could be emptied efficiently for the glow peaks to be observed, τ_{rec} should be sufficiently smaller than the time taken in recording the glow peak fully. Let us call this as the time span T_R of the glow curve. How small τ_{rec} should be in comparison to T_R is discussed in Sect. 5.7. Low value of τ_{rec} is achieved if the values of $n(T)$, $m(T)$, M, A_h, and A_m are high. For $n(T)$ to have high value, n_0 should be high. The effect of n_0 value is clearly seen in the OTOR model which is dealt in the following section. Let us consider the value of τ_{rec} in set F of Lewandowski and Mckeever [7]. In that set LM use $N = M{=}10^9$ m^{-3} and

$A_h = R(T)v(T) \sim = 10^{-17}$ m^{3}s^{-1} so that the value of τ_{rec} is about 10^8 s. So the free carriers thrown up by thermal excitation cannot relax into the recombination centers until the time elapses far beyond 10^8 s (refer Fig. 4 in [7]). Similar condition is now seen in modified set A shown in Fig. 5.2. As expected, the TL emission is extremely low and falls very slowly after reaching the maximum. On the other hand in modified set F, the results of which are shown in Fig. 5.3, the value of τ_{rec} is decreased to about 10 s by increasing the values of N and M to a higher value of 10^{16} m^{-3}. The values of $P(T)$ remain unaltered. In Fig. 5.4 are plotted the values of τ_{rec} as a function of T for the original and the modified sets A and F. It may be seen that like the TL glow curves and the $Q(T)$ values, the profiles of τ_{rec} also are reversed, that is, τ_{rec} of sets A (modified) and F (modified) are like those of original sets F and A, respectively.

The above results show that the important requirement for emptying the traps efficiently and thereby obtaining a proper glow peak is that τ_{rec} should be sufficiently smaller than the glow peak time span. In our Fig. 5.3 described above, the value of τ_{rec} is about two orders of magnitude smaller than the time span T_R of the glow peaks. This range of τ_{rec}/T_R ratio provides sufficient time for emptying the traps with the heating rate of 4 K/min used in these experiments. The reason for the high values of τ_{rec} in original set F and modified set A may be clearly understood from (5.9a–c) by substituting the values of n, M, and A_h of these sets into this equation. Obviously, the low values of these parameters in these sets are responsible for the high value of τ_{rec}.

5.3.2 More Examples of Dependence of QE on System Parameters

In addition to the NMTS model dealt above, Sunta et al. [8] have studied the effect of parameter values on QE for two other physically meaningful models namely the OTOR and the IMTS (refer Sect. 3.3.2). In the OTOR model, the energy level scheme used is same as in the GG model [2] (see Sect. 3.3.2) which assumes $A_n = A_h =$ constant A. In GG model R_{ret} is always greater than R_{rec} except when the traps are filled up to saturation. The glow curves are calculated for given sets of input parameters both by using the analytical (3.15) as well as by using the rate Eqs. (3.22)–(3.24). The results from the two are compared by plotting them on the same graph. The profiles of $Q(T)$ and $P(T)$ functions are also calculated. For GG model these functions are

$$Q(T) = [(ns\exp(-E/KT) - n_cA(N-n)]/n_cAn \tag{5.10}$$

$$P(T) = (N-n)/n \tag{5.11}$$

Figures 5.5a, b, and c show the results. The input parameter values are included in the respective figure caption. Figure 5.5a shows that in glow curves 1 and 2

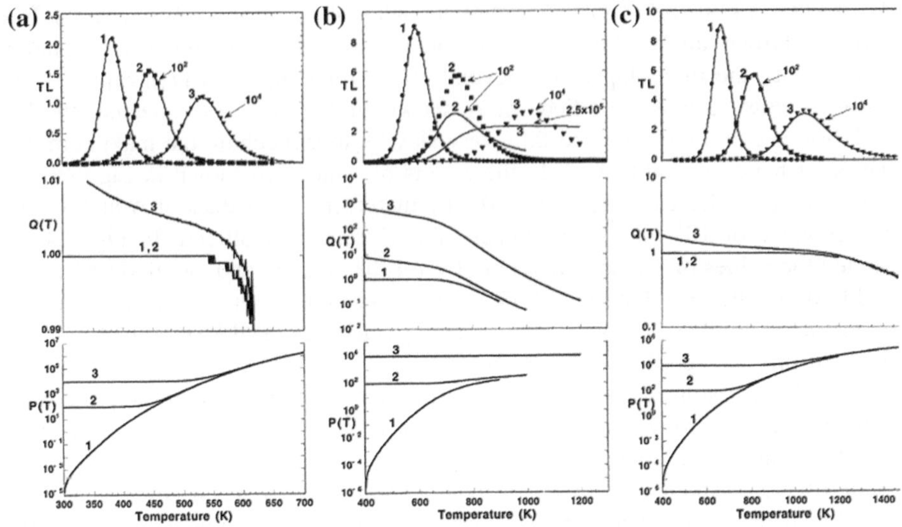

Fig. 5.5 TL glow curves, $Q(T)$ and $P(T)$ profiles showing the effect of change in N, n_o and σ values for the GG model. Input parameters which are common in a, b and c are $E = 1$ eV, $\beta = 1$ K s^{-1}. Curves 1, 2 and 3 in each of them are for $n_O/N = 1$, 10^{-2} and 10^{-4} respectively. The continuous line curves are from numerical calculations using the rate (3.22), (3.23) and (3.24) in which $A_n = A_h = \sigma v$. Values shown with •, Δ, ∏ are from analytical (3.15). Numbers shown in each glow peak are the factors by which the actual intensity values are multiplied for plotting. **a** $N = 10^{18}$ m^{-3} and $\sigma = 10^{-18}$ m^2 and s $= 10^{12}$ s^{-1} In **b** $N = 10^{18}$ m^{-3} and $\sigma = 10^{-23}$ m^2 and s $= 10^7$ s^{-1}. In **c** $N = 10^{21}$ m^{-3} and $\sigma = 10^{-23}$ m^2 and s $= 10^7$ s^{-1}. Sunta et al. [8, Figs. 1, 2 and 3] .

analytically and numerically calculated glow curves coincide with each other, while in curve 3 the numerical values are slightly smaller than the analytical values. This is because in the case of curve 3 the n values are much smaller than N, which reflects in the $Q(T)$ value departing from 1 by up to 1 %. The $P(T)$ values range from 10^4 to 10^6 in this case. We see that in these cases QE is well satisfied even when we have $P(T) \gg 1$. In Fig. 5.5b the input value of parameter A is decreased by a factor of 10^5, other parameters remaining same as in Fig. 5.5a. This results in the analytical glow curves 2 and 3 becoming way apart from the numerical ones. As expected we have $Q(T) \gg 1$ and $P(T) \ll 1$ as in Fig. 5.5a. In Fig. 5.5c the input value of N is increased by a factor of 10^3, other parameters remaining same as in Fig. 5.5b. The glow curves as well as the $Q(T)$ and the $P(T)$ profiles are now similar to those in Fig. 5.5a. It is thus seen that the behavior of the glow curves as well as that of the $Q(T)$ and the $P(T)$ functions depend on the values of N, n_O and A in this model and the changes in the values of $Q(T)$ are unrelated to those in the values of $P(T)$. In other words the functions $Q(T)$ and $P(T)$ possess their values independent of each other and that $P(T) \ll 1$ is not essential to achieve QE. It is only the absolute values of the parameters namely N, n_0, M, A_h, and A_n as discussed above, that determine whether the QE condition would be satisfied. The examples solved above show that greater are the values of these parameters, smaller is the life time τ_{rec}, of the thermally liberated carriers in the conduction band. A low value of τ_{rec} disallows the free carriers n_c, to buildup in the conduction band and thus helps to achieve the QE condition ($n_c \ll n$ and $|dn_c/t| \ll |dn/dt)|$.

Another physically meaningful model is the IMTS model (see Sect. 3.5.2). As stated before, this model may be considered to be nearest to the case of real materials which have multiple types of traps at deeper as well as shallower levels with reference to the AT. It is important to remember that in IMTS model, the TDDT when filled up, produce additional RC population (to balance the charge) and when empty, they recapture the charge carriers liberated from the AT. The recapture may be considered as nonradiative recombination in the rate equations of this model. Thus the TDDT population helps in the relaxation of excited carriers irrespective of whether the TDDT are filled up or empty. Therefore, the denominator in (5.1) and (5.2) should be modified to $R_{rec} + R_{recap}$, for the case of this model, where R_{recap} stands for the rate of recapture by the empty TDDTs. The $Q(T)$, $P(T)$, and the τ_{rec} in the IMTS model thus are

$$Q(T) = [(ns\exp(-E/kT) - n_c(N-n)\sigma_n v)]/[(n_c(n+m)\sigma_h v + n_c(M-m)\sigma_m v)] \tag{5.12}$$

$$P(T) = (N-n)\sigma_n/[(n+m)\sigma_h + (M-m)\sigma_m] \tag{5.13}$$

$$\tau_{rec} = 1/[(n+m)A_h + (M-m)A_m] \tag{5.14}$$

where, as stated before, $A_h = \sigma_h\ v\ and\ A_m = \sigma_m\ v$.

At $n \ll N$, M and $m \ll M$, one would have $P(T) \approx (N\sigma_n/M\sigma_m)$. Thus one may get $P(T) \geq 1$ or ≤ 1 depending on the relative values of $N\sigma_n$ and $M\sigma_m$. However, in this model we expect $M \gg N$, since in real phosphors there usually are several TDDTs at a number of deeper levels. The total concentration of traps at these various levels is included in the parameter M. We would therefore normally expect $M\sigma_m \gg N\sigma_n$ and thus we would have $P(T) \ll 1$ in this model. Under the QE condition (subject to $M\sigma_m$ and $N\sigma_n$ values being sufficiently high) the condition $M\sigma_m \gg N\sigma_n$ produces glow peaks of FO kinetics (see Sect. 3.5.2).

5.4 Lower Limit of Parameter Values for the Validity of QE Assumption

In the preceding sections, it is shown that greater are the values of the parameters N, M, n_0, A_n, and A_m better it is for the QE to be satisfied. This statement describes the condition for QE in qualitative terms. In this section, we discuss the conditions for satisfying the QE quantitatively and make an attempt to find the lower limit of these parameters above which QE would always be satisfied [12]. Under the QE condition, the rate of relaxation of the free carriers from the conduction band should be equal to the excitation rate. This may be expressed as

$$R_{ex} = R_{rec} + R_{ret} \tag{5.15}$$

Let us now consider the NMTS and the IMTS models which may plausibly be applicable to the real materials. Let us first consider the NMTS model with low occupancy of the active traps ($n \ll N, M$). According to (5.15)for satisfying the QE condition we should then have

$$n(N_c\sigma_n v)\exp(-E/kT)/n_c[N\,\sigma_n v + M\,\sigma_h v] = 1 \tag{5.16}$$

We have appropriately considered the trap occupancy to be low, since when QE is satisfied at low trap occupancies it would always be satisfied at higher trap occupancies too in the glow curves (see for example Figs. 5.5 and 5.6 and compare $Q(T)$ values at low and high trap occupancies). Note that in (5.16), we have used $(N_c\sigma_n\ v)$ in the numerator in place of the frequency factor s (see Appendix-1 and Sect. 3.2.2). The first term in the denominator on left hand side of this equation represents R_{ret} and the second term represents R_{rec}. The functions $\sigma_n v$ and $\sigma_h v$ are used for A_n and A_h, respectively. (See Appendix 1). This equation leads to

$$n_c/n = N_c\sigma_n \exp(-E/kT)/[N\,\sigma_n + M\,\sigma_h] \tag{5.17}$$

Similarly for IMTS model, at low trap occupancies ($n \ll N$, $m \ll M$) we should have for QE condition

$$n(N_c\sigma_n v)\exp(-E/kT)\ /\ n_c[N\,\sigma_n v\ +\ M\,\sigma_m v] = 1. \tag{5.18}$$

As in the case of NMTS model, this leads to

$$n_c/n = N_c\sigma_n \exp(-E/kT)/[N\,\sigma_n + M\,\sigma_m] \tag{5.19}$$

In this equation, we see that the function in the denominator provides only for retrapping in vacant active traps and capture (trapping) in vacant TDDTs. This is because the recombination rate is negligible, i e., $(n + m)\sigma_h \ll (N\,\sigma_n + M\,\sigma_m)$ at low trap occupancies in IMTS model. It does not, however, mean that there would be no recombination and no TL, since sufficient number of active traps, n and RC, $(n + m)$ exist even when the condition $(n + m)\sigma_h \ll (N\,\sigma_n + M\,\sigma_m)$ is holding. From (5.17) and (5.19), we see that to satisfy $n_c \ll n$ and $dn_c/dt \approx 0$, we should have high values for $N\,\sigma_n$ and $M\,\sigma_h$ in NMTS and for $N\,\sigma_n$ and $M\,\sigma_m$ in IMTS. If we designate N and M as N_i and σ_n, σ_h and σ_m as σ_i, we may state that QE would depend critically on $\Sigma_i N_i\sigma_i$. Sunta et al. [8] computed the glow curves and the profiles of QE functions namely the ratios n_c/n and $[|dn_c/dt|]/[|dn/dt|]$ for OTOR, $n_c/(n + M)$ and $|dn_c/dt|/|dn/dt|$ for NMTS and $n_c/(n + m)$ and $[|dn_c/dt|]/[|-dn/dt-dm/dt|]$ for IMTS model, using the rate equations of the respective models and selected sets of parameter values for each of these models. The parameter values in

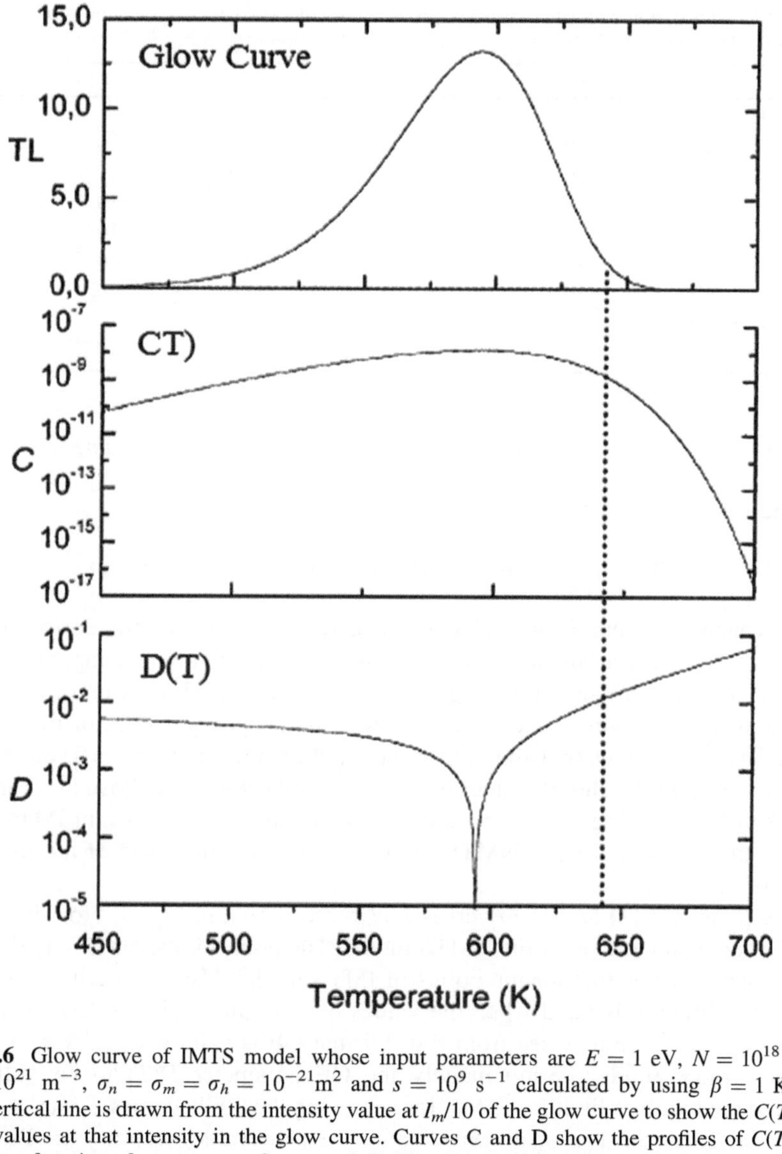

Fig. 5.6 Glow curve of IMTS model whose input parameters are $E = 1$ eV, $N = 10^{18}$ m^{-3}, $M = 10^{21}$ m^{-3}, $\sigma_n = \sigma_m = \sigma_h = 10^{-21}$m^2 and $s = 10^9$ s^{-1} calculated by using $\beta = 1$ K s^{-1}. The vertical line is drawn from the intensity value at I_m/10 of the glow curve to show the $C(T)$ and $D(T)$ values at that intensity in the glow curve. Curves C and D show the profiles of $C(T)$ and $D(T)$ as a function of temperature. Sunta et al. [8, Fig. 6b]. © IOP publishing Ltd. Reproduced by permission. All rights reserved

each set of IMTS and NMTS models and the corresponding values of the above QE parameters in these profiles at 1/10 of the glow peak maximum (so as to use the values at low trap occupancy) on the fall side of the glow peak are listed in Table 5.1. The QE functions described above are abbreviated as $C(T)$ and $D(T)$. For IMTS model

Table 5.1 Input parameter values for six glow curves of IMTS model and one glow curve of NMTS model, used for plotting *C(T)* and *D(T)* as a function of $\Sigma_i N_i \sigma_i$ (m^{-1}) in Fig. 5.7

N	M	σ_i	$\Sigma_i N_i \sigma_i$	$C(T)$	$D(T)$	τ_m
IMTS glow curves						
10^{18}	10^{18}	10^{-23}	2×10^{-5}	10^{-2}	10	1
10^{18}	10^{19}	10^{-23}	1.1×10^{-4}	10^{-4}	1	10^{-1}
10^{18}	10^{20}	10^{-23}	10^{-3}	10^{-4}	10^{-1}	10^{-2}
10^{18}	10^{21}	10^{-23}	10^{-2}	10^{-6}	10^{-2}	10^{-3}
10^{18}	10^{18}	10^{-19}	2×10^{-1}	10^{-6}	10^{-3}	10^{-4}
10^{18}	10^{18}	10^{-18}	2	10^{-7}	10^{-4}	10^{-5}
NMTS glow curves						
10^{18}	10^{16}	10^{-18}	1	–	10^{-4}	10^{-3}

$$C(T) = n_c(T)/\{n(T) + m(T)\} \text{ and } D(T) = |\mathrm{d}n_c/\mathrm{d}t|/|-\mathrm{d}n/\mathrm{d}t - \mathrm{d}m/\mathrm{d}t| \quad (5.20)$$

And for NMTS model

$$\mathrm{C(T)} = n_c(T)/[n(T) + M] \text{ and } D(T) = |\mathrm{d}n_c/\mathrm{d}t|/|\mathrm{d}n/\mathrm{d}t \quad (5.21)$$

A typical glow curve of IMTS model and the profiles of the corresponding *C(T)* and *D(T)* functions are shown in Fig. 5.6 with common temperature axis scale. A vertical line drawn through 1/10 of the glow peak maximum meets the *C(T)* and *D(T)* profiles at the required low trap occupancy values of these functions. The glow curves (not shown) for the input parameters listed in Table 5.1 are used for plotting the chosen values of *C(T)* and *D(T)* described above as a function of $\Sigma_i N_i \sigma_i$ (m^{-1}) in Fig. 5.7. The parameter σ_i stands for σ_n and σ_m in IMTS glow curves and for σ_n and σ_h in NMTS glow curves. Units of N and M are m^{-3} and those of σ_i are $m^{\ 2}$.

The plots of *C(T)* and *D(T)* versus $\Sigma_i N_i \sigma_i$ are shown in Fig. 5.7 for six sets of IMTS model and one set of the NMTS model. The point for the NMTS model falls on the same line as that for the points of IMTS model. Using the criteria that the QE approximation is valid when the values of *C(T)* and *D(T)* are less than 10^{-2} (refer Sect. 5.1), we may see from Fig. 5.7 that QE is valid when $\Sigma_i N_i \sigma_i \geq 10^{-2}$. When a more rigid criterion namely the QE parameter *D(T)* to be $\leq 10^{-3}$,is adopted the required limit for $\Sigma_i N_i \sigma_i$ for achieving the QE becomes $\Sigma_i N_i \sigma_i = 10^{-1}$. However, from the practical consideration of achievable accuracy in TL measurements it appears appropriate to set the limit as $\Sigma_i N_i \sigma_i \geq 10^{-2}$ (see Sect. 5.1). At low concentration of active traps ($n_0/N < 10^{-2}$), it is only the deep traps *M*, which provide for the RC. They act as radiative RC in NMTS model and as nonradiative RC in IMTS model. Thus the quantity $M\sigma_i v$ is the minimum value of the recombination probability. This means the maximum value of the recombination life time τ_m for the free electrons of the conduction band is given by

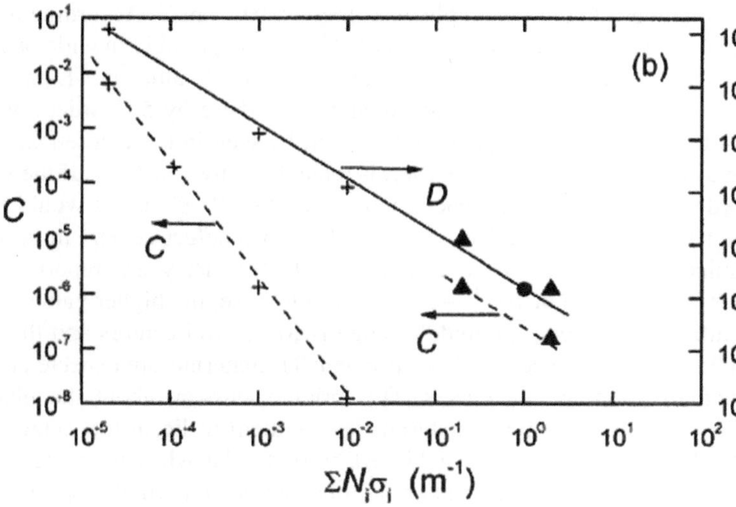

Fig. 5.7 $C(T)$ and $D(T)$ values at I_m /10 plotted as a function of $\Sigma_i N_i \sigma_i (\mathrm{m}^{-1})$ for six sets of parameters of IMTS and oneset of NMTS. The input parameter values of individual sets with the $\Sigma_i N_i \sigma_i$ values and the corresponding $C(T)$ and $D(T)$ values are listed Table 5.1. Sunta et al. [8, Fig. 7b].

$$\tau_m = 1/(M\sigma_i v) \tag{5.22}$$

5.5 Validity of QE Assumption in the TL of Inorganic Phosphors

As stated before, Kelly et al. [6] had, on the basis of theoretical calculations, concluded that QE assumption is valid when the concentration N of the active traps is more than 10^{15} cm^{-3}. In the above discussion, it is shown that QE depends not only on the value of concentration of the traps but it actually depends on the product of concentration and the cross-section of the traps. The above theoretical conclusions highlighting the role of N, M, and the cross-sections σ_n and σ_m in attaining the QE condition may be used to answer the question whether the QE condition prevails in the TL emission process of various phosphors and thereby whether the analytical expressions based on the QE approximation can legitimately be used to analyze the experimental glow curves. The answer hinges directly on the range of N, M, and σ values that may be applicable to these materials. TL phosphors in general have dopant concentrations as high as hundreds to thousands of ppm, which implies that the TL-related defect center concentrations should be of a similar order. The F center concentration in divalent cation doped alkali halides [13], for example, is as high as 10^{17} cm^{-3}. According to Rose [14] defect centre concentration as high as 10^{19} cm^{-3} is possible in wide band gap materials. Accordingly, one may expect that the *N* and *M* values in TL phosphors may be of similar order. Even a value smaller than these by 5–7 orders of magnitude is sufficiently large for QE to hold as can be seen in the theoretical results given above. The σ value as stated earlier, is related to s by $s = N_c\sigma v$. If the *s* value is of the order of the Debye frequency (approx. 10^{12}–10^{13} s^{-1}), one would have a σ value of 10^{-12}–10^{-13} cm^2. The σ values of various defect centers in impurity-doped semiconductors have been estimated [14, 15]. They are reported to be spread over a wide range of 10^{-12}–10^{-16} cm^2. Of these, the higher values are for Coulomb attractive centers, the middle range is for neutral centers and the lower range pertains to repulsive centers [14]. Efficient TL materials are expected to have traps with high trapping efficiencies. This brief discussion about the plausible values of *N*, *M*, and σ in doped materials shows them to lie in the range within which the QE condition is readily achieved, in particular when the material concerned is a wide band gap TL phosphor. TL phosphors are normally optimized for high trap concentrations by doping with impurities and annealing procedures. Thus if, for example, the density of the active and the deep traps (*N* and *M* values) is as low as 10^{12} cm^{-3},a σ value as low as 10^{-16} cm^2 may be sufficient to reach the critical limit of $\Sigma_i N_i \sigma_i = 10^{-2}$ m^{-1} to satisfy the QE condition. As discussed above, the values of *N*, *M* and σ are known to be much greater than these. This discussion shows that the pessimism expressed by Kelly et al. [6] regarding the validity of the QE approximation for analyzing the experimental glow peaks is not justified and that the use of analytical expressions which are derived by using the QE approximation are quite legitimate for real phosphors.

5.6 Dependence of QE on Heating Rate

Sunta et al. [9] computed TL glow curves using different values of heating rate β for four sets of parameters (see Table 5.2) of IMTS and model using the respective rate equations (see Sect. 3.5.2). They also computed the corresponding profiles of the QE functions namely *Q*(*T*) and *D*(*T*). The profiles of the TL glow together with those of the *D*(*T*) and *Q*(*T*) for set 1 and set 4 of parameter values listed in Table 5.2 are given in Figs. 5.8 and 5.9. These figures show that the values of *D*(*T*) as well as *Q*(*T*)increase progressively as β is increased. Looking at the *D*(*T*) profiles one may say that the glow curves for whom the *D*(*T*) values exceed 10^{-2} do not satisfy the QE condition. These results show that smaller is the value of *D*(*T*), closer to equilibrium ($dn_c/dt \approx 0$) is the system. It is observed that simultaneous with the increase in *D*(*T*), the glow peaks become progressively wider and their shape factor δ/ω begins to increase when β is increased beyond a certain limit. At such high heating rates, the δ/ω value exceeds 0.52. It is seen that this limit depends on the values of

Table 5.2 Parameter values used in calculating the glow curves and the $D(T)$ and the $Q(T)$ profiles at different heating rates for four parameter sets

Case (set) number	N	M	n_o	m_o	A_n	A_m	A_h	s	τ_m
1	10^{18}	10^{17}	10^{16}	10^{15}	10^{-18}	10^{-18}	10^{-18}	10^{7}	10
2	10^{18}	10^{19}	10^{16}	10^{17}	10^{-18}	10^{-19}	10^{-19}	10^{7}	1
3	10^{18}	10^{18}	10^{16}	10^{16}	10^{-13}	5×10^{-18}	5×10^{-18}	10^{12}	0.2
4	10^{18}	10^{22}	10^{16}	10^{20}	10^{-18}	10^{-18}	10^{-18}	10^{7}	10^{-4}

The values of N, M, n_o, and m_o are in m^{-3}, those of A_n, A_m, and A_h are in $m^3\ s^{-1}$, s *in* s^{-1}. The activation energy E in all the sets is 1 eV. The maximum values of recombination life time τ_m (s) also is listed

the input parameters. The changes in $D(T)$ are seen prominently in Fig. 5.8 and not in Fig. 5.9. This is because the minimum value of β to manifest the change in δ/ω in the latter cases is much greater than the highest value of β used in these figures. This value in its turn depends on the input parameter values. It is important to understand why the β value influences the QE condition and how the latter in its turn influences the δ/ω value of the TL glow peak as β is increased. An increase in β causes a shift of the glow peak to higher temperature. An increase in temperature causes a rapid increase in the excitation rate R_{ex}, since R_{ex} increases exponentially (for a given value of n) with the sample temperature T (see (5.3)). The probability of clearance (given by R_{rec}) of the charge carriers n_c, from the excited state on the other hand remains the same. This is because the radiative and nonradiative recombination rates depend on the concentration and cross-section of RCs and the vacancies in TDDTs which are not influenced by the increase in sample temperature. In fact there may be a minor decrease in cross-sections, because of their $T^{-3/2}$ dependence [7], with increase in temperature which may cause a small decrease in the clearance rate. As the heating rate increases, the imbalance between the excitation rate and the relaxation rate increases progressively. At a given heating rate the effect of this imbalance is clearly seen in the slow fall of the glow curve intensity at $T > T_m$. This leads to the widening of the glow curve and increasing the value of its symmetry factor δ/ω. The heating rate at which this effect sets in depends on the values of the parameters N, M, σ_n, σ_h, and σ_m on whom depends also the relaxation rate of the free carriers n_c, from the excited state. This may be more clearly understood by comparing Figs 5.8 and 5.9. It may be seen that in the latter figure the parameter values are much greater than in the former set (see Table 5.2). In the case of the former figure there occurs a buildup of free carriers n_c, in the excited state at high heating rates. This effect is clearly seen in Fig. 5.10 in which n_c/n and R_{ex}/R_{rec} are plotted as a function of T during the heating phase for the case of two glow curves computed with $\beta = 0.001$ and $\beta = 10\ Ks^{-1}$ shown in Fig. 5.8 as glow curves 1 and 5 respectively. It is seen (Fig. 5.10) that at $\beta = 0.001\ Ks^{-1}$, ratio $n_c/n = 10^{-5}$–10^{-3} and $R_{ex}/R_{rec} = 1$ which clearly is a perfect QE condition like in RW model. On the other hand at $\beta = 10\ Ks^{-1}$, the ratio n_c/n becomes as high as 10^2 at T_m and goes beyond 10^8 on the fall side of the glow curve. The value

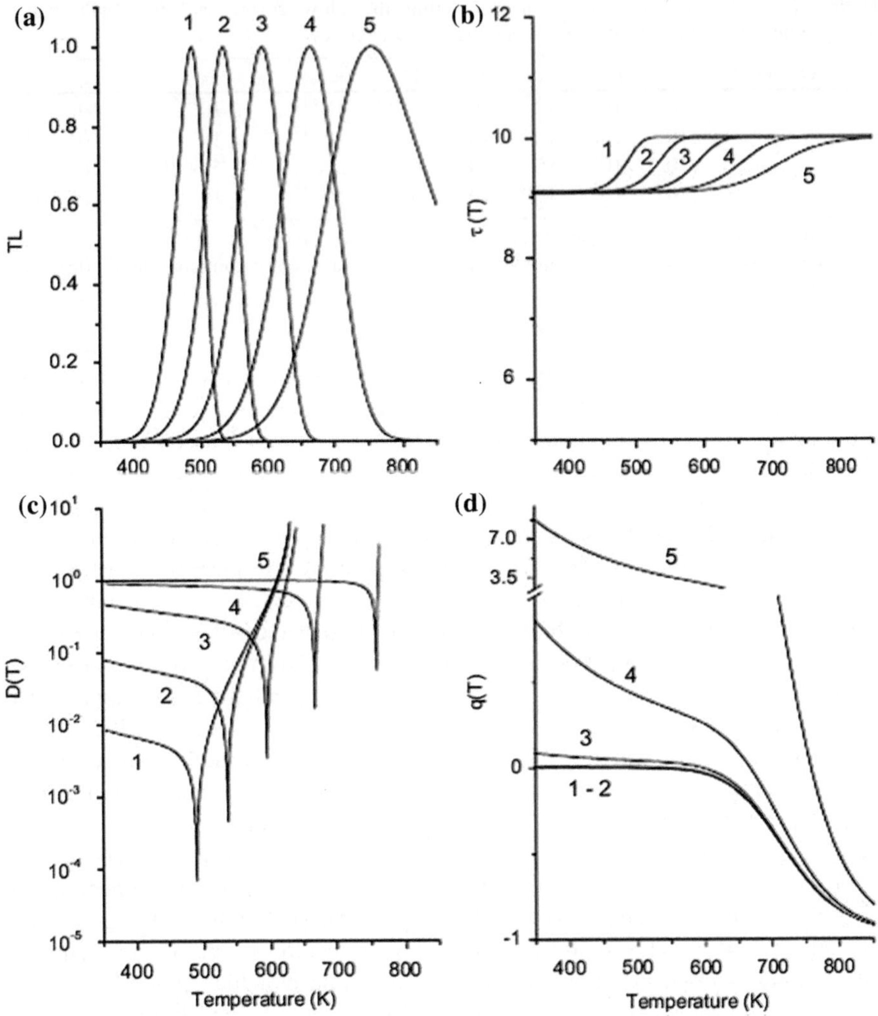

Glow curve number					
	1	2	3	4	5
$\beta(Ks^{-1})$	0.001	0.01	0.1	1	10
δ/ω	0.425	0.426	0.436	0.465	0.594
T_R(s)	139000	16400	2005	279.5	78

◀ **Fig. 5.8** **a** TL glow curves for input parameter values, $E = 1$ eV, $N = 10^{18}$ m^{-3}, $M = 10^{17}$ m^{-3}, $n_0 = 10^{16}$ m^{-3}, $m_0 = 10^{15}$ m^{-3}, $A_n = A_m = A_h = 10^{-18}$ m^3 s^{-1}, s = 10^7 s^{-1}. Glow curves 1 to 5 are for heating rates 0.001, 0.01, 0.01, 1 and 10 Ks^{-1} respectively. Peak values are normalized to 1 for all the 5 glow curves. The shape factors δ/ω and TR for glow curves 1 to 5 respectively are as given in table above for different heating rates. **b** Recombination life time as a function of *T* along the glow curve. **c** *D*(*T*) values as a function of *T* along the glow curve. **d** *q*(*T*) values (where *q*(*T*) = *Q*(*T*)-1) as a function of *T* along the glow curve. Sunta et al. [9, Fig. 4]

of $R_{ex}/R_{rec} = 10^2$ on the rising part of the glow curve and falls to $<10^{-5}$ toward the end of the glow curve. These values of n_c/n and R_{ex}/R_{rec} at $\beta = 10$ Ks^{-1} reflect a strong disequilibrium condition in contrast to the perfect QE at $\beta = 0.001$ Ks^{-1}. Thus we see that because of higher excitation rate and inefficient clearance of the charge carriers from the excited state, the TL glow curve becomes wider. The widening effect is more pronounced on the fall side half of the glow curve. This is because the fall in intensity in this part becomes progressively slower with increasing *T*. This happens because the active trap population *n* gets progressively depleted which in its turn leads to predominance of retrapping over recombination ($R_{ret} \gg R_{rec}$ see (5.4) and (5.5a–c) with $n \ll N$). This causes the untrapped charge carriers to recycle through a sequence of retrapping and untrapping before their final relaxation into the RCs leading to the emptying of the trap at a slow pace. The declining part of the glow curve in this situation may appear as a long tail. While discussing this behavior of the simulated glow curves Chen and Pagonis [11] also give example of some experimental samples which appear with long tailed glow curves. These include studies in mica by Kristianpoller et al. [16] and Sakurai [17], in $CaSO_4$: Dy by Mathur et al. [18], in $M_2Si_5N_8$: Eu (where M = Ca, Sr, Ba) by Smet et al. [19] and in BaCl: OH^- by Furetta et al. [20]. However, it may not be appropriate to compare the result of simulation in this case with that of the experimental samples, because it is unlikely that the long tailed glow peaks of experimental samples could be due to disequilibrium condition in the conduction band, which as explained above, takes place in simulated glow curves at high temperatures. Further, as discussed in the preceding section, one expects QE condition to be satisfied in the TL of the experimental samples mentioned above. The reason for the long tails in the experimental glow curves may be due to the presence of the low intensity accompanying peaks on the higher temperature side of the main peak. One may be able to separate them by heating the irradiated sample at different temperatures and recording the TL after each heating step.

Figures 5.8 and 5.9 show that the *D*(*T*) values are much more sensitive to changes in β than the *Q*(*T*) values at $D(T) \ll 1$ which is the range of interest to examine the closeness of the system to QE. It may be seen in Fig. 5.8 that when *D*(*T*) changes from $<10^{-2}$ to about 5×10^{-1}, the *Q*(*T*) value remains at about 1 only. At high heating rates when the system is far from QE, the *D*(*T*) attains a constant value equal to 1 ($dn_c/dt \approx dn/dt$) at $T < T_m$, but it increases rapidly at $T > T_m$. As a result of its high sensitivity to changes in β the profiles of

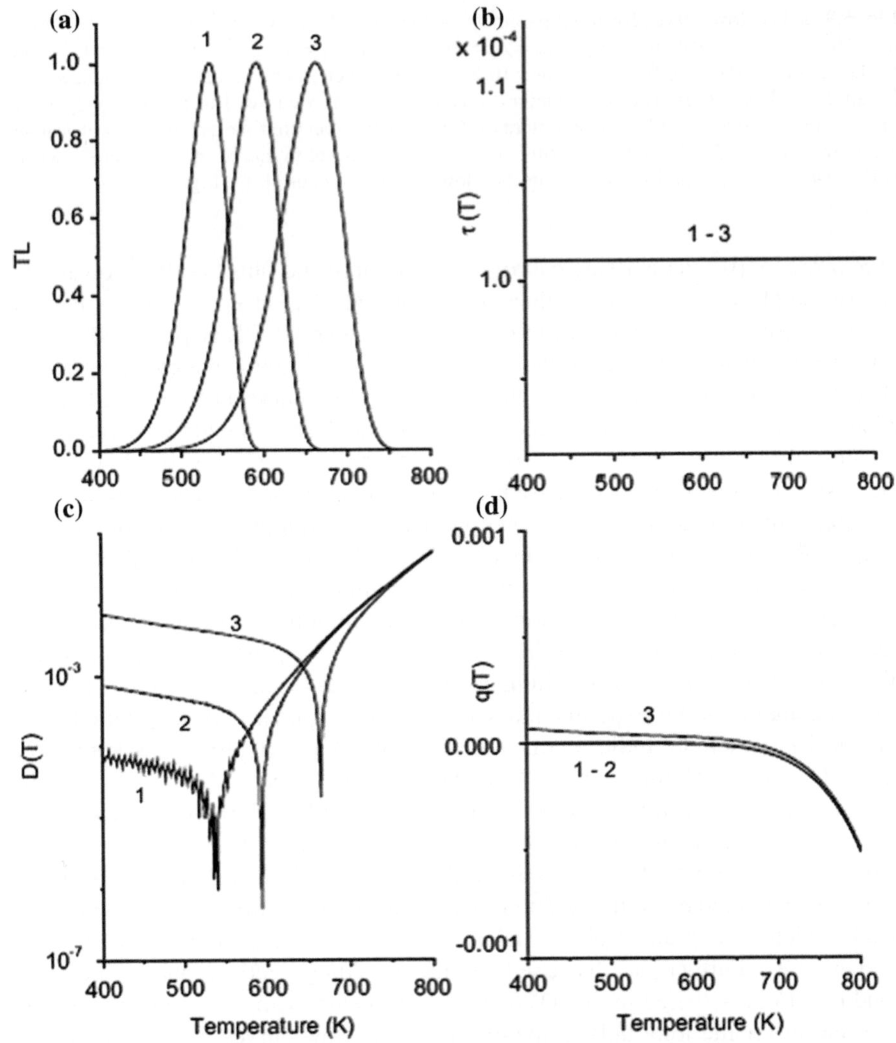

Glow peak number			
	1	2	3
β (Ks^{-1})	0.1	1	10
δ/ω	0.426	0.423	0.425
T_R(s)	1612.5	194	24

◀ **Fig. 5.9** **a** TL glow curves for input parameter values, $E = 1$ eV, $N = 10^{18}$ m^{-3}, $M = 10^{22}$ m^{-3}, $n_0 = 10^{16}$ m^{-3}, $m_0 = 10^{20}$ m^{-3}, $A_n = A_m = A_h = 10^{-18}$ m^3 s^{-1}, s $=10^7$ s^{-1}. Glow curves 1 to 3 are for heating rate = 0.1, 1 and 10 Ks^{-1} respectively. Peak values are normalized to 1 for the 3 glow curves. The shape factors of all the three glow curves are nearly 0.425. The time span of glow curves 1, 2 and 3 respectively are as given in table above. **b** Recombination life time as a function of T along the glow curve. **c** $D(T)$ values as a function of T along the glow curve. **d** $q(T)$ values (where $q(T) = Q(T)$-1) as a function of T along the glow curve. Sunta et al. [9, Fig. 5]

$D(T)$ clearly show how far or how close to QE a system is when the glow curve is being recorded using a given heating rate β.

5.7 Changes in the Glow Peak Time Span T_R and the Glow Peak Shape Due to the Change in the Heating Rate

The glow peak time span T_R may be defined as $T_R = (T_e - T_0)/\beta$, where T_0 and T_e, respectively, are the temperatures at the beginning and the end of the glow curve. If the sample is sensitive enough T_e and T_0 may be taken at $I_m/100$ on either side of the peak maximum. As expected, T_R decreases with increase in β. For each glow peak, the value of the recombination life time τ (see (5.9a, b, c) and (5.12)) also is calculated. Its maximum value τ_m also is listed in Table 5.2 for each set of parameters used in the computations. The plots of δ/ω and T_R/τ_m are shown together as a function of β in Fig. 5.11. It is seen as said before, that the δ/ω values begin to increase after the β value crosses a certain limit. As may be seen this limit is different for different parameter sets. Notwithstanding these differences, it emerges from these plots that for all the four parameter sets, the δ/ω values take an upward trend in δ/ω versus β curves at approximately same value of T_R/τ_m. Figure 5.11 shows that the upward trend appears at approximately $T_R/\tau_m = 10^3$. It means that QE is no more satisfied when $T_R/\tau_m \leq 10^3$ in IMTS model. These values of T_R/τ_m seem to be only weakly influenced by the input parameters used. The IMTS model being realistically related to the plausible energy level scheme of the real materials, we may adopt the $T_R/\tau_m = 10^3$ as the lowest limit of the T_R/τ_m ratio above which the QE approximation would be valid in any system of parameters of a TL phosphor. It may be noted in Fig. 5.8 that the values of δ/ω are not so sensitive to the changes in β as those of $D(T)$. For example, we may see in this figure that when β is increased from 10^{-3} to 10^{-2} K s^{-1}, the $D(T)$ value increases from about 7×10^{-3} to about 7×10^{-2}, whereas a change in the value of δ/ω is seen only when β exceeds 6.67×10^{-2} K s^{-1}. This means that the glow peak has already entered the non-QE range at $\beta < 10^{-2}$ K s^{-1}, whereas a perceptible increase in δ/ω is seen only at $\beta = 6.67 \times 10^{-2}$ K s^{-1}. This means that the vertical dotted lines tentatively drawn by the authors [9] in Fig. 5.11 to indicate the transition of the glow curves from QE into non-QE range may need to be shifted to lower β side by about one decade. This may push the corresponding

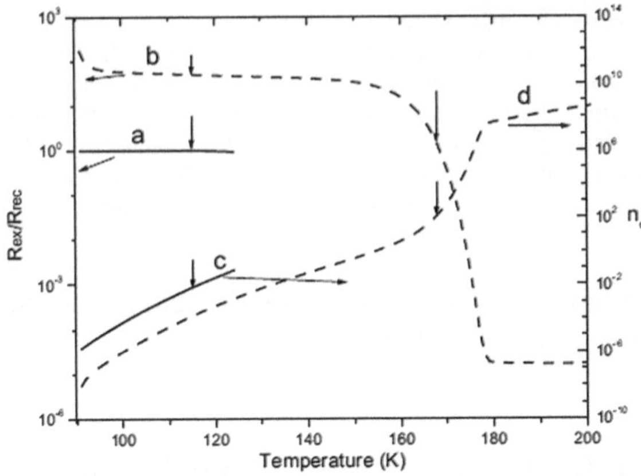

Fig. 5.10 R_{ex}/R_{rec} (*left Y* scale) and n_c/n (*right Y* scale) at two heating rates (β) for the glow curves of an NMTS case. Curves a and c are for $\beta = 0.001\ \text{K s}^{-1}$ and curves b and d are for $\beta = 10\ \text{K s}^{-1}$. *Vertical arrows* are marked at $T = T_m$ for the glow peaks of $\beta = 0.001\ \text{K s}^{-1}$ and $\beta = 10\ \text{K s}^{-1}$ respectively. Sunta et al. [9, Fig. 6].

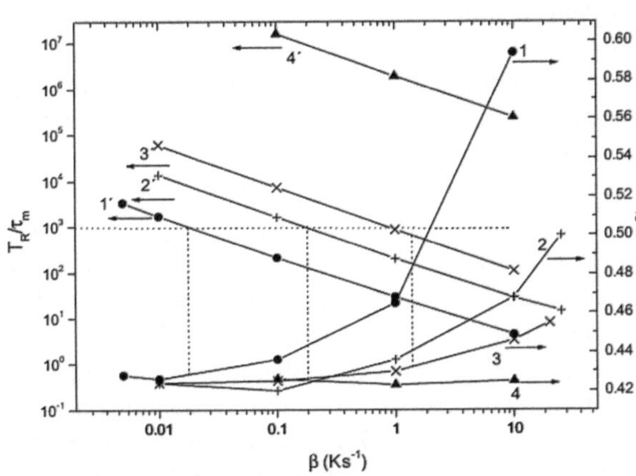

Fig. 5.11 Plots of δ/ω (*right Y* scale) and T_R/τ_m (*left Y* scale) as a function of β for four cases of IMTS model (see Table 5.2) indicated respectively as curves 1–4 for δ/ω versus β and curves $1'$ to $4'$ for T_R/τ_m versus β curves. *Vertical dotted lines* are drawn to show that the points at which the δ/ω curves for cases 1–3 begin to take an upward direction correspond to $T_R/\tau_m \approx 10^3$ for all the three cases. For case 4 the upward deviation in δ/ω curve is not seen, since the β value for this trend would appear beyond the range of β value plotted here Sunta et al. [9, Fig. 9].

T_R/τ_m values further up by about one decade. This means that the limit of T_R/τ_m value above which a glow curve may safely be considered to meet the QE condition, may be put at 10^4. Notwithstanding the approximate nature of this criterion for the QE approximation, the property of the heating rate dependence of QE and the peak shape may be usefully applied to verify whether an experimental glow peak under study is being recorded under the QE condition at the heating rate being used. Further one may even attempt to assess the parameter values of the given phosphor material. This is discussed in the following.

5.8 Test for QE for the Glow Peaks of Real Materials

The quantities $D(T)$ and $Q(T)$ whose numerical values tell us how far or how close to QE the system is during TL readout, are not measurable for the glow curves of the real samples. McKeever et al. [10] show that the shape of the $q(T)$ curve where $q(T) = Q(T)-1$ may be found by using the simultaneously measured TL and TSC glow curves of the sample. Actual values of the $q(T)$, however, cannot be found due to the lack of the knowledge of the scaling parameters. Sunta et al. [9] developed a simple method based on the change of the shape of the glow curve by changing the heating rate β. As discussed above, the glow peaks become broader in width as the heating rate β is increased. At the same time, the values of the QE parameter $D(T)$ and $Q(T)$ also increase with increase in β. At certain stage of high value of β the value of the $D(T)$ becomes greater than 10^{-2} and that of $Q(T)$ exceeds 1, which means that at this value of β the glow curve goes into the non-QE condition. Further increase in β leads to a clearly visible change in the shape of the glow curve as well. The glow curve set in which such effects are seen is given in Fig. 5.8. On the other hand, no such effect is seen in Fig. 5.9. In this latter set of the glow peaks there is no change in the glow peak shape factor δ/ω even at heating rates as high as 10–20 K s^{-1}. Obviously, in this case the values of the input parameters N, M and the cross-sections of the corresponding traps are high enough and therefore the QE conditions are well satisfied (refer Sect. 5.4) even up to such high heating rates. The heating rate dependence of the glow curve shape under non-QE conditions provides a simple and unambiguous method to test whether the QE approximation is satisfied during the read out of any glow curve. To check whether the QE is holding when a TL is being recoded at a given heating rate β, Sunta et al. [9] suggest that the experimenter should repeat the glow curve recording at two more heating rates β_1 and β_2 which may preferably differ from the given β by an order of magnitude on lower and higher side, respectively. If the δ/ω value remains same, it means that the system is in QE at all the three heating rates. If δ/ω increases with β_2 but does not decreases with β_1, it means that β is near the limiting values of heating rate above which QE may not be satisfied. On the other hand if δ/ω decreases with β_1, it may mean that QE is not satisfied at heating rate β. In such a case, one has to reduce the heating rate appropriately, so that the QE condition is satisfied during the glow curve recording for the given sample. This means that while making a statement regarding the validity or otherwise of the QE condition one should also state the β value at which the TL measurements have been carried out to analyze a glow peak. Sunta et al. [9] applied this method to test the QE in the case of the well-known TLD phosphor LiF-TLD-100. The glow peak No. 5 of this phosphor was recorded at three β values namely 1, 5, and 25 K s^{-1}. The glow peak shape factor δ/ω remained unaltered by these changes in β. This means that QE is well satisfied even at such a high heating rate as 25 K s^{-1} for this glow peak of LiF-TLD phosphor.

5.9 Assessing the Free Carrier Life Time and the Parameters of a Real Sample

As seen above, the value of the ratio T_R/τ_m at which the δ/ω value takes an upward turn (transition from QE to non-QE condition), appears approximately at 10^3 irrespective of the parameter values of the system. The near constant value of T_R/τ_m at the turning point may be used to assess the maximum value of the life time τ_m for an experimental glow curve of the sample under study. For this purpose, the experimenter has to record the glow curve at various linear heating rates until the δ/ω versus β curve takes an upward trend. One may go up to as high a heating rate as experimentally feasible. The sample should be irradiated with a dose well below saturation so that the condition $n \ll N$ and $m \ll M$ holds. Under this condition, we get $\tau_m = 1/MA_m$ (where $A_m = \sigma_m v$ in IMTS model). Assuming that the free charge carrier is an electron, we may use $v = 10^5$ ms^{-1}. Substituting $1/MA_m$ for τ_m, in $T_R/\tau_m \approx 10^3$, we arrive at

$$M\sigma_m \approx 10^{-2}/T_R \tag{5.23}$$

This relation provides a means for finding the approximate value of $M\sigma_m$ for any given experimental sample, provided, of course that the sample shows a change in glow peak shape with change in β. If one of the parameters, M or σ_m is known or is found by some other technique, the other can be estimated by using the measured value of T_R. In case no change is observed in peak shape even at the highest usable value of β (see for example, no change in glow peak shape is observed in Fig. 5.9 in which highest value of β is 100 Ks^{-1}), one would then be able to say that the value of $M\sigma_m$ for the phosphor under investigation is greater than $10^{-2}/T_R$ where T_R is the glow curve time span measured at the highest value of β used in the experiment. It is to be noted here that though the M and σ_m values in (5.15) are for TDDT and do not include those for AT, this relation provides a good indicator of the range of the TL trap concentrations and their cross-sections in general. Let us once again take the example of glow peak 5 of the well-known TLD phosphor LiF (TLD-100), in whose case, as described above, the T_R values at the heating rates 1, 5 and 25 K s^{-1} are found to be 108, 23 and 4.7 s, respectively.

No change is found in the peak shape up to the heating rate of 25 K s^{-1}. It means that T_R is expected to be less than 4.7 s in order to apply (5.23), which in other words means that we have $M\sigma_m > 2 \times 10^{-3}$ for this glow peak. The parameter M which represents the concentration of TDDT may be related to the Z_3 type centers in this material (i.e., perturbed F-type centers (see Fig. 2.1)) (see Caldas et al. [21] and Landreth and Mckeever [22]. The concentration of this type of centers in impurity-doped alkali halides is known to lie in the range of 10^{22}–10^{23} m^{-3} (see Sonder and Sibly [13]). Assuming $\sigma_m = 10^{-16}$–10^{-20} m^2 (see Rose [14] and Lax [15]), one would expect $M\sigma_m > 10^2$ m^{-1} for the above-mentioned Z_3 type centers of LiF-TLD-100. This means that τ_m ($\approx 1/M\sigma_m v$) may be less than 10^{-7} s. This in its turn means that the value of T_R at the β value at which δ/ω takes an upward turn would be less than 10^{-4} s. The values of T_R and β are grossly related according to $\beta = k/T_R$ with the constant k varying very slowly and having a gross value around 10^2 (see the above stated results on LiF(TLD-100) and also Munish Kumar et al. [23] on α-Al_2O_3 and the simulated glow curves). Using this relation, we would expect the β value for the upward turning point of δ/ω versus β curve (see Fig. 5.11) to be around 10^6 K s^{-1}. These values of T_R and β tell us that in the real phosphors the possibility of a non-QE condition may be ruled out at any reasonably employable heating rate to record their TL glow curves. The example of LiF (TLD-100) may be considered as a typical case of the impurity-doped inorganic insulating synthetic materials and minerals used in dosimetry and TL dating.

References

1. J.T. Randall, M.H.F. Wilkins, Proc. Roy. Soc. (London) series **A184**, 365 (1945)
2. G.F.J. Garlick, A.F. Gibson, Proc. Phys. Soc. (London) **62**, 574 (1948)
3. C.E. May, J.A. partridge, J. Chem. Phys. **40**, 1401 (1964)
4. R. Visocekas, La Luminescence de la calcite après irradiation cathodique, TL et luminescence par effettunel, Ph.D. Thesis UniversitePiere et Marie Curie, Paris (1978)
5. D. Shenker, R. Chen, J. Comput. Phys. **10**, 272 (1972)
6. P. Kelly, M.J. Laubitz, P. Braunlich, Phys. Rev. **B4**, 1960 (1971)
7. A.C. Lewandowski, S.W.S. Mckeever, Phys. Rev. **B43**, 8163 (1991)
8. C.M. Sunta, W.E.F. Ayta, R.N. Kulkarni, J.F.D. Chubaci, S. Watanabe, J. Phys. D Appl. Phys. **32**, 717–725 (1999)
9. C.M. Sunta, W.E.F. Ayta, J.F.D. Chubaci, S. Watanabe, J. Phys. D Appl. Phys. **34**, 3285–3295 (2001)
10. S.W.S. McKeever, B.G. Markey, A.C. Lewandowki, Nucl. Tracks Radiat. Meas. **21**, 57 (1993)
11. R. Chen, V. Pagonis, J. Lumiescence **143**, 734–740 (2013)
12. C.M. Sunta, W.E.F. Ayta, J.F.D. Chubaci, S. Watanabe, Radiat. Meas. **35**, 595 (2002)
13. E. Sonder, W.A. Sibley, in *Point defects In solids*, ed. by J.F. Crawford, L.M. Slifkin (Plenum, New York, 1972), p. 201
14. A. Rose, *Concepts in Photoconductivity and Allied Problems* (Interscience, New York, 1963), p. 118
15. M. Lax, Phys. Rev. **119**, 1520 (1960)

16. N. Kristianpoller, Y. Kirsh, S. Shoval, D. Weiss, R. Chen, Nucl. Traks Radiat. Meas. **14**, 101 (1988)
17. T. Sakurai, J. Appl. Phys. **82**, 5722 (1997)
18. V.K. Mathur, A.C. Lewandowski, N.A. Guardala, J.L. price, Radiat. Meas. **30**, 735 (1999)
19. P.F. Smet, K. Eekhout, A.J.J. Bos, E. Van der Kolk, P. dorenbos, J. Lumin. **132**, 682 (2012)
20. C. Furetta, M.T. Laudadio, C. Sanipoli, A. Scacco, J.M. Gomez-Ros, V. Correcher, J. Phys. Chem. Solids **60**, 957 (1999)
21. L.V.E. Caldas, M.R. Mayhugh, T.G. Stoebe, J. Appl. Phys. **54**, 3431 (1983)
22. J.L. Landreth, S.W.S. Mckeever, J. Phys. D Appl. Phys. **18**, 1919 (1985)
23. M. Kumar, G. Chaurasiya, R.K. Kher, B.C. Bhatt, C.M. Sunta, Ind J. Pure Appl. **47**, 402 (2009)

Chapter 6
Intensity Growth with Dose

Abstract This chapter deals essentially with the nonlinear growth of thermoluminescence (TL) intensity with radiation dose. After a brief review of earlier theories and earlier works on this subject, the discussion takes up the so-called heating stage competition model to explain the nonlinear behavior of TL growth with dose. As against the arbitrary competitors assumed by earlier workers, in the proposed model the competitors are thermally disconnected deep traps (TDDT) which when empty act as competitors and when filled cause increase in the number of recombination centers (RC). When the active traps (AT) and TDDTs get filled up during irradiation, the reduction in competition and the increase in filled active trap population takes place simultaneously. As a result, the TL intensity growth becomes superlinear. The treatment of the model takes into account the irradiation and heating stages together. Using this model, expressions are derived for the TL intensity growth curve, supralinearity factor (SF), and predose sensitization factor (PDSF). Also derived is a new expression called the sensitization factor (S_nF). The factor S_nF which has been introduced by this author is used to give a unified explanation for the mechanisms that give rise to the *SF* and the *PDSF*. The computed profiles of *SF*, *PDSF*, and S_nF are compared with the experimentally obtained profiles of these factors for the case of LiF:Mg, Ti TLD phosphor. Apart from providing a theoretical basis underlying the phenomena of supralinearity and predose sensitization, the model dispels the doubt of some workers that the mechanism involved in these two phenomena may be altogether different from each other.

6.1 Introduction

One would normally expect that thermoluminescence (TL) intensity should grow linearly with dose. This normal behavior is seen in many mineral samples [1] and in a minority of well-known TLD phosphors which include Al_2O_3:C [2], LiF:Mg,

C. M. Sunta, *Unraveling Thermoluminescence*,
Springer Series in Materials Science 202,
DOI: 10.1007/978-81-322-1940-8_6, © Springer India 2015

Cu, P, and $Li_2B_4O_7$:Cu [3]. In most other materials used in radiation dosimetry, different degrees of deviation from linearly are observed. Srivastava [3] has presented a comparative study (see Fig. 6.1) of the response factors of various TLD phosphors as a function of gamma radiation dose. It may be seen from the figure that the response factor is uniformly the same (normalized to unity) at dose levels smaller than about 1 Gy for all phosphors. In this dose range, TL intensity increases linearly with dose. At doses greater than 1 Gy the response factor increases verily with dose for the different phosphors shown in the figure except two of them. To the list of the six phosphors showing increased response in Fig. 6.1 may be added the TL phosphor based on mineral CaF_2 [4] and Mg_2SiO_4:Tb [5]. Finally, at high doses the response factor falls after going through a maximum for all of them. This fall is essentially due to saturation in the growth of TL intensity with further increase in dose. The deviations in response factors of individual materials have been found to depend on various factors. Those in particular are the type of radiation (UV, β, γ, X-rays, or heavy charged particles such as protons and alphas), energy of radiation, temperature of the glow peak [6], pre-irradiation treatment, impurity content, and heating rate [7–9]. While studying the UV dose response of the TL of semiconducting diamonds, Halperin and Chen [6] found that exposure to UV rays of wavelength 300–400 nm produced a strong superlinear dose response proportional to D^2 to D^3 (where D is the dose), whereas exposure to 225 nm produced a linear response with dose. The authors explained that the multistage transition of electrons from valence band to conduction band is responsible for superlinearity in the case of lower energy UV photons. They state that high energy photons of 225 nm (5.5 eV) help the electrons to transit from the valence band directly into the conduction band. This produces a linear growth of TL intensity with UV exposure. Strong superlinearity is observed also in 110 °C peak of synthetic quartz irradiated by beta or gamma rays [10]. The explanation for this behavior has been given on the basis of competition between the radiative and nonradiative recombinations during heating to read the TL glow. Superlinearity is eliminated if the sample is heated at a high temperature before irradiation. It is believed that pre-irradiation heating destroys the competitors responsible for nonradiative recombination. The concept of competitors is elaborated later in this chapter. The superlinear behavior of most TL dosimetry materials has been explained on the basis of the competition theory. The above stated case of semiconducting diamond is among a few exceptions where the competition theory is not applicable and where the pattern of dose response is related to the TL excitation stage.

A material that has attracted much attention of TL workers is the popular LiF:Mg, Ti (TLD-100) phosphor used in radiation dosimetry. The TL growth pattern of this material, when irradiated with gamma or beta rays, is shown in Fig. 6.2. This may be described as (1) linear for doses lesser than 10 Gy, (2) superlinear in the dose range 10–200 Gy, and (3) sublinear at doses 200–2,000 Gy followed by saturation at doses above 2,000 Gy. Similar type of growth pattern (linear, superlinear, sublinear, saturation) is seen in most TL phosphors as discussed above. While explaining the mechanism of nonlinearity, Lakshmanan et. al.

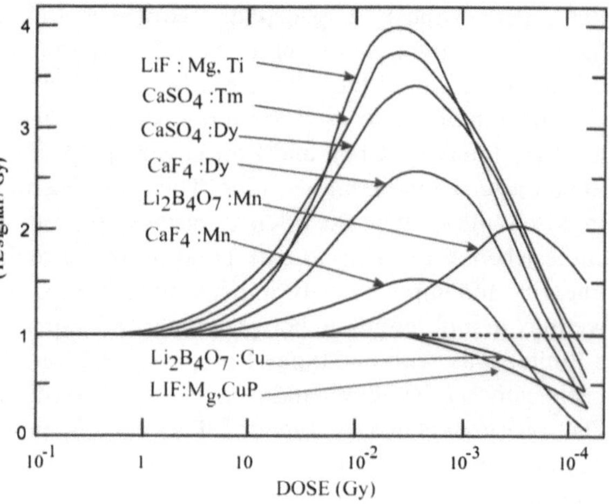

Fig. 6.1 TL intensity per Gy of gamma ray dose (normalized response factor) for different TLD phosphors. (From Srivastava [3])

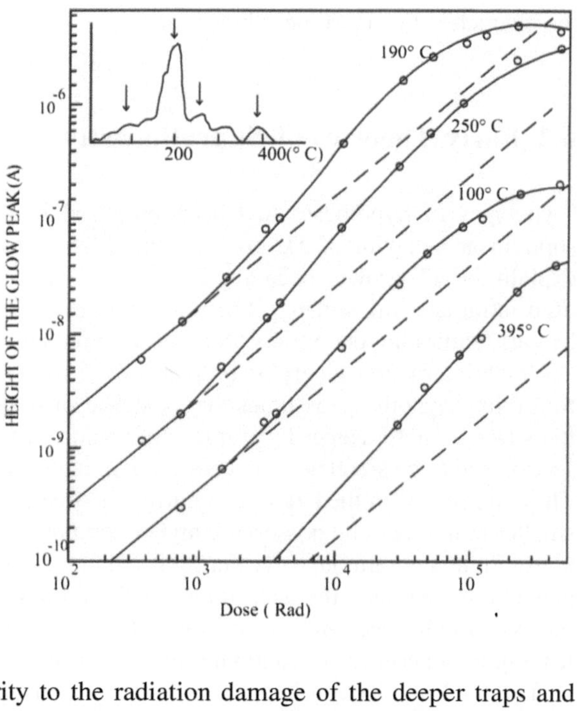

Fig. 6.2 TL intensity as a function of dose for different glow peaks of LiF:Mg, Ti (TLD-100). The glow curve is shown as *inset*. The glow peaks corresponding to the TL intensity growth curves are marked with *arrows* (Author's unpublished work). Rad, the old unit of dose is equal to 1/100 of the presently used unit Gy

[11] attribute the superlinearity to the radiation damage of the deeper traps and Bhuwan Chandra et al. [12] correlate the sensitization to the residual TL which also is due to deeper traps. Srivastva and Supe [13] assume a quasi-continuous competing trap to explain the nonlinear response. As we will see later in this chapter, both superlinearity and sensitization are caused by filling of the deeper traps and not by damage to them. The superlinear part is seen to follow the initial

linear part in most TL phosphors. However, this may not be a rule. Aitken [14] found that in the case of quartz extracts from pottery samples the superlinear part precedes the linear part. To describe this kind of TL intensity growth behavior various authors have used the terms supralinearity, superlinearity, or the general term nonlinearity. Chen and McKeever [15] have discussed quantitative methods to characterize the nonlinearities. They propose the use of the term superlinearity in dose ranges where the intensity growth is more than linear and supralinearity in cases where correction in extrapolation is the main issue. They define the superlinearity and supralinearity indices mathematically. To put in a simpler language we may first differentiate between the terms super and supra. Super is concerned in a limited way with one type of property which in the present case is linearity. The term supra on the other hand is an inclusive term. For example, the Oxford English Dictionary explains supranational as having power and influence that extends beyond national boundaries. The term supralinear appears appropriate for TL intensity growth curves which have both linear and superlinear regions irrespective of which region precedes and which follows. The individual linear and superlinear parts may be referred as such. The following pages are devoted to a discussion of the characteristic supralinear behavior.

6.2 Early Theories of Supralinearity

Two types of hypotheses had been proposed in the late 1960s to explain the supralinear behavior of TL growth. These hypotheses were used in particular to explain the TL growth behavior of the main TL peak of LiF:Mg, Ti (TLD-100) used in radiation dosimetry. These hypotheses are: (1) competition in trap filling by early saturating deep traps proposed by Suntharalingam and Cameron [16] and (2) track interaction theory as proposed by Claffy et al. [17]. The competition in trap filling hypothesis assumes the existence of traps at levels deeper than active traps (AT). These deeper level traps are presumed to possess higher capture cross section and to be smaller in concentration than the AT of the glow peak of interest. They compete with the latter in capturing the free carriers during irradiation. Being smaller in number and possessing higher capture cross section they get filled up at a faster rate and saturate early during irradiation. The competition faced by the AT gets eliminated once the deep traps get filled up. Consequently the filling rate of the AT is enhanced. As a result the AT produce higher TL yield per unit dose, leading to superlinearity following the initial linear growth.

The track interaction hypothesis assumes that the incident radiations produce tracks of ionization. At smaller doses the tracks are far enough from each other (inter-track distances more than the range of released charged carriers during the heating). At higher doses the inter-track distance decreases and during heating some of the released carriers from one track are able to travel to the neighboring track. These undergo recombination with the RC of the neighboring tracks, thus leading to an increased TL yield. At smaller doses the released carriers from any one track are able to produce TL only by recombination within the same track. Any of the carriers escaping from the track are assumed to the captured nonradiatively in the inter-track region. The track interaction hypothesis was originally proposed irrespective of the type of radiation. However, in the case of low LET (linear energy transfer) radiations such as X, gamma, and beta rays of energy greater than 500 keV, the track interaction theory is not likely to be applicable, since the products of ionization in this case are nearly isotropically distributed in space rather than forming isolated tracks. However, the heavy charge particles (HCP), such as the alpha and protons are expected to produce the tracks which are nearly of cylindrical geometry. For example a 4 MeV alpha particle deposits 90 % of dose within 100 μm and 98 % within 200 μm of the center of the track [18]. Obviously, within the tracks the concentration of filled traps and RC would be high, whereas in the space outside the tracks there would be empty trapping sites which capture the released carriers from the tracks nonradiatively during heating. At higher doses of HCP irradiation the inter-track distance would decrease. The thermally released carriers would then have a chance to travel from one track to the other leading to an increase in the probability of radiative recombination. This causes TL intensity as a function of dose to rise more than linearly as the HCP dose is increased. A proof of this model was provided by Rosenkrantz and Horowritz [19] who demonstrated that supralinearity in 4 MeV alpha irradiation increases strikingly and begins at lower doses when the incident alpha particles fluence changes from near parallel to near isotropic geometry, which effectively reduces the inter-track distance.

The deep trap competition hypothesis was originally proposed particularly to explain the supralinear behavior of the main glow peak of LiF:Mg, Ti (TLD-100), which appears at about 200 °C (the exact temperature depends on the heating rate). The hypothesis had the weakness that no experimental evidence was available to substantiate the existence of deep traps and their early saturating property. The glow curves, using the then available systems could be measured only up to about 300 °C. A direct way to show the existence of the traps at levels deeper than that of the glow peak of interest could be to detect glow peaks at temperatures greater than that of the glow peak under study. Later, Sunta and co-workers designed and fabricated such a glow curve reader system in their laboratory which is described briefly in [20]. The first set of the glow curves read with this system up to a temperature of 600 °C is reported by Sunta [21] for the blue-green mineral CaF_2 TLD. Using this system to study the TL of LiF (TLD-100) Sunta et al. [8] recorded five more glow peaks at temperatures greater than that of the main glow peak of LiF:Mg, Ti (TLD-100), the phosphor which had then been the subject of investigation. The results show that the high temperature glow peaks (appearing from 200 to 500 °C) do not saturate at the dose level at which superlinearity begins for the glow peak of 200 °C. In fact the glow peaks from the deeper traps (those with T_m in 200–500 °C range) seem to saturate at temperatures higher than the saturation temperature of the above stated 200 °C peak (see Fig. 6.2). Notwithstanding this, it may not be said that competition in trap filling stage may be totally absent.

However, Chen and Fogel [22] using numerical calculations found that the contribution of the trap filling stage competition in causing supralinearity is far smaller than that of heating stage competition. The role of heating stage in supralinearity is well supported by the fact that supralinearity factor (SF) is found to depend on the heating rate used during TL readout [7]. The models proposed for heating stage competition are discussed in the following.

6.3 Competition During Heating Stage

The hypothesis of heating stage competition was proposed originally by Rodine and Land [23] to qualitatively explain the quadratic TL growth behavior of some of the glow peaks of beta irradiated ThO_2. Subsequent authors have given this hypothesis a mathematical footing [7, 24–26]. The approach of these authors has the common theme that a nonradiative capture of the released charge carrier from the active trap competes with its radiative recombination during heating. The competitor may be a trap or even a recombination center. This scheme is shown in Fig. 6.3. Kristianpoller et al. [24] assume the competitor to be a trap. The scheme is essentially the same as described under the IMTS model (see Sect. 3.5.2) in Chap. 3). The rate equations to describe the transport of released charge carriers during the TL read out heating also are the same. Kristianpoller arrives at the expression for the area S of the glow curve as

$$S = [A_n/(A_m M)]h_0 n_0 \tag{6.1}$$

where A_n and A_m are the capture coefficient of the AT and the competitors, respectively, and M is the total concentration of the competitors. h_0 and n_0, respectively, are the initial concentration of the RC and the AT. This expression shows that the TL intensity would increase quadratically with dose if both h_0 and n_0 increase linearly with dose. To arrive at this conclusion Kristianpoller et al. [24] make the following arbitrary assumptions in course of the mathematical treatment: (1) retrapping in AT is negligible in comparison to thermal excitation rate like in RW model, (2) the filled active trap population n_0 is negligibly small in comparison to the vacancies in the competing traps ($n_0 \ll (M - m_0)$), (3) the competing traps are nearly empty when the heating is started ($m_0 \ll M$), and the capture rate by them is much faster than both the retrapping and the radiative recombination rates combined ($A_m (M - m) \gg [A_n(N - n) + A_h h]$, where A_h is the recombination coefficient and h is the concentration of the RC). It may be noted that in this scheme the TL growth is quadratic right from the beginning of the intensity growth curve.

Chen and co-workers [27] have discussed a model consisting again of one active trap but two RC, transition into one of them being nonradiative, like in the case of the competitor trap. Superficially, this model may appear identical to that

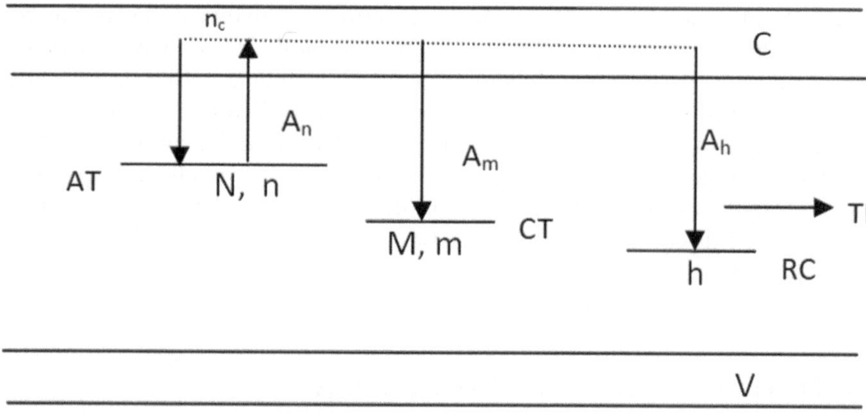

Fig. 6.3 Scheme of competition during TL read out heating as used by Kristianpoller et al. [24]. AT = active trap, CT = competing trap, RC = recombination center, N and M are the total concentration and n and m the filled concentrations of AT and CT, respectively. A_n, A_m, and A_h are the probabilities of transition of the free carriers n_c into the AT, CT and RC, respectively. Note that the notations used here are different from those used by Kristianpoller et al. to maintain the uniformity of notations in the book

of Kristianpoller et al. [24]. However, if we consider the charge balance conditions, the total number of trapped carriers (including those in competitor traps) should balance with the total number of RCs. We would then have $n_1 + n_2 = h$ in the competitor trap case and $n = h_1 + h_2$ in case of competitor RC. Here n and n_1 stand for the filled active trap population, n_2 stands for the filled competitor trap population, and h and h_1 stand for radiative RC population and h_2 that for non-radiative RC. The mathematical treatment in the two cases would obviously be different and the results would depend on the relative values of parameters chosen (for example the values of ratios n_1/n_2, h_1/h_2, A_{n1}/A_{n2}, A_{h1}/A_{h2} etc.). Results for a selected set of parameters for the case of RC as competitor are given by Chen et.al. [27]. In contrast to the competing trap case, in which Kristianpoller et al. [24] get quadratic TL growth all through, in the competing RC case the initial growth is seen to be linear followed by moderate superlinearity before approaching the saturation. This behavior seems to match with that of some of the TL phosphors. The superlinearity seems to begin near the dose range where the competing RC saturates. This seems to be the result of choosing smaller value for the competing RC population in comparison to that of radiative RC, when the capture cross sections for the released carrier during heating for both types of RC are taken to be equal. Obviously, the results from the numerical computations depend on the relative as well as the absolute values of the parameters chosen. Chen and McKeever [28] suggest that both kinds of competitions (during excitation as well as during heating) may contribute to the observed supralinearity in a material, although they state that the governing effect is that of competition during heating. Dealing with the combined effect of competition during excitation and readout stage of TL, Lee and Chen [29] solved the equations of the two stages sequentially using the QE assumption and found the expression for the TL yield S numerically. They found that if recapture predominates over recombination a quadratic dose dependence results. If however, the competitor trap is filled up to near saturation level, the value of S is seen to increase linearly with dose. In the following sections both the excitation and the heating stages are dealt analytically producing the results not only for the extreme cases but dealing with all levels of trap filling. The results are discussed in the perspective of the experimental results in many of the TL phosphors.

6.4 Thermally Disconnected Deep Traps as Heating Stage Competitors

Sunta et al. [25] incorporated some meaningful changes in the heating stage competition model. They gave a physical meaning to the competitor trap by assuming that it is a thermally disconnected deep trap (TDDT). Essentially in this case the authors are dealing with the IMTS model, in which case the charge carriers released from the AT during heating may be recaptured in the TDDT. The behavior of glow curves for this model has been discussed in Chap. 3 (see Sect. 3.5.2).

Sunta et al. [25] introduced the trap filling functions for the AT and the TDDTs which act as competitor traps. They calculated the TL intensity growth profiles as a function of dose for different ratios between the total population N of AT and M that of the TDDT. The theoretical treatment proceeds as follows: If the trap filling rate dn is taken to be directly proportional to dose dD and also directly proportional to the vacancies in the traps (this is commensurate with the linear growth of optical absorption (OA) versus dose for the traps and the centers in a variety of phosphors), we may write that

$$dn = \alpha(N - n)dD \tag{6.2}$$

$$dm = \varepsilon(M - m)dD \tag{6.3}$$

where α and ε are the fraction of the vacant traps that get filled per unit dose. The meaning of the quantities N, n, M and m is already given in Chap. 3 as well as given in the Appendix. Using the condition $n = 0$ and $m = 0$ at $D = 0$, the above equations, respectively, lead to

$$n = N[1 - \exp(-\alpha D)] \tag{6.4}$$

$$m = M[1 - \exp(-\varepsilon D)] \tag{6.5}$$

where n and m are the concentrations of the filled AT and the filled TDDT on giving dose D and N and M are their total concentrations. The coefficients α and ε being much smaller than 1, the values of n and m rise linearly with dose D over many decades, till αD and $\varepsilon D \ll 1$. Thus we get $n = N\alpha D$ and $m = M\varepsilon D$ at doses which are much below the saturation dose. The values of n and m become sub-linear before reaching the saturation level. At saturation we get $\alpha D \approx 1$ in (6.4) which means $n \approx N$. Similarly at $\varepsilon D \approx 1$ we would have $m \approx M$. The difference between saturation doses for them depends on the relative values of α and ε. If $\alpha = \varepsilon$ both types of traps reach saturation at the same dose. If $\alpha < \varepsilon$, the value of n will reach saturation later than m and vice versa. During heating when a charge carrier is released from the active trap and is delocalized it has three options for relaxation: (1) to get retrapped into vacant sites of ATs, (2) to recombine radiatively with an RC, and (3) to get recaptured nonradiatively into the vacant sites of TDDT which means the TDDTs function as competitors to radiative recombination in this scheme. The effect of option 1 is only to delay the final emptying of the AT, since ultimately all the trapped charge carriers n, find their way either into RCs or into vacant sites of TDDTs; both these being thermally stable at the temperature at which ATs are being emptied. Return of charge carriers from these entities is not permitted during the heating phase. The variables n and m represent the number of trapped carriers in the ATs and in the TDDTs, respectively, after receiving the given dose D. Let us assume that the overall number out of the total number of charge carriers n that falls into RCs and produces TL, is given by the equation [25, 26]

$$I = n\left[\frac{(n+m)A_h}{(n+m)A_h + (M-m)A_m}\right] \tag{6.6}$$

The parameters A_h and A_m have the usual meaning as described in Chap. 3 (refer also Appendix).The quantities $(n + m)$ and $(M - m)$ are the available number of RCs and the number of vacancies in TDDTs, respectively, after the given dose D. The quantity I is the integrated TL intensity (the area of the glow peak which is designated by S in some of the expressions). The quantity within the square brackets is the fraction of trapped carrier population, n which recombines radiatively and is called the radiative recombination factor F. Equation (6.6) assumes that F remains constant throughout the TL readout heating and is equal to its value at the start of heating. This obviously is an approximation, since n and m vary with temperature during the heating to read out the TL, whereas in the above equation these parameters represent their values at the start of heating. However, Sunta et al. [26, 30] show that this assumption turns out to be generally valid for the parametric conditions encountered in practical cases. This may be understood from the following: In (6.6) use $[(n + m)A_h]/[(M - m)A_m] = R(T)$, where $R(T)$ is a function of temperature T. The function $F(T)$ within the square brackets in (6.6), defined earlier as the recombining fraction may then be written as

$$F(T) = R(T)/[1 + R(T)] \tag{6.7}$$

Equation (6.7) shows that $F(T)$ remains constant during the readout heating if $R(T)$ is constant or $R(T) \gg 1$. The latter case which means $F(T) \approx 1$ at all T, implies that it is a recombination dominant case. Constancy of $R(T)$ follows from the result that both $(n + m)$ and $(M - m)$ are found to decrease in about the same proportion during heating in most of the cases. Using numerical computations for a wide range of parametric combinations Sunta et al. [26, 30] show the variety of conditions under which $R(T)$ is found to remain constant or its variation is insignificant. Even in the cases in which $R(T)$ is seen to vary significantly with T, its effect on the value of $F(T)$ averaged over the full glow curve is found to be insignificant. An exceptional case in which the average value of $F(T)$ differs significantly from its value at T_0 occurs when $R(T)$ changes from <1 to ≫1 during heating, which means it is a case in which recapture by TDDTs dominates in the beginning of heating phase but further rise in temperature turns it into recombination dominant. This happens due to the competing TDDTs getting filled up by recapture of the untrapped electrons from the AT. This in its turn is found to happen when the AT as well as the TDDTs are filled to an intermediate level (1/10 < filling level < saturation) and $A_h \ll A_m \ll A_n$. Such a combination of parameters may, however be quite uncommon. Secondly, we are in general concerned with trap filling levels which are much below saturation.

Let us use the simplifying assumptions $M = aN$, $A_m = A_h$ and $\varepsilon = \alpha$ in (6.6) and substitute (6.4) and (6.5) for n and m, respectively. We then get

$$I = \frac{N(1 + a)(1 - e^{-\alpha D})^2}{(a + 1 - e^{-\alpha D})} \tag{6.8}$$

For doses well below saturation, we would have $\alpha D \ll 1$. The above equation then reduces to

$$I = \frac{(\alpha D)^2 N(a + 1)}{(a + \alpha D)} \tag{6.9}$$

At low doses, one expects $\alpha D \ll a$ (highest value of αD is about 1, which is attained only near the saturation dose, that is at $n \approx N$). Equation (6.9) thus shows that at sub-saturation doses, the TL intensity grows quadratically with dose D. Figure 6.4 shows the plots of (6.8) for different values of a. It may be seen that all the curves from 1 to 4 show quadratic dose dependence, although the dose range to satisfy the quadratic behavior goes down progressively when the value of a decreases, which means the smaller the number of TDDTs in comparison to the ATs, the lower the dose range in which the quadratic behavior is seen. In such cases the growth profile at higher doses may be seen to be less than quadratic (see for example curves 3 and 4 in Fig. 6.4). This is because in such cases the condition $\alpha D \ll a$ may be satisfied at such low doses which are well below the range of this

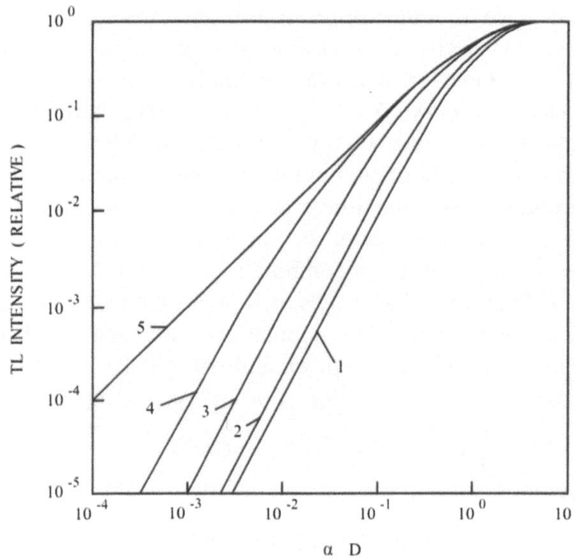

Fig. 6.4 TL intensity versus dose for the interactive trap system. *Curves 1–5* are for $a = 10, 1, 0.1, 0.01$ and 0, respectively. The *X*-axis is marked in terms of αD. The TL intensity is normalized to unity at saturation [25, Fig. 2].

computation. All the growth curves ultimately saturate at the same level of dose, i.e., at $\alpha D \approx 1$. Although schematically (see Fig. 6.3) the present scheme appears similar to that of Kristianpollar, we see that apart from giving a physical meaning to the competitor, in the present scheme we get the full profile of TL growth as a function of dose till saturation. Further, the growth pattern depends on the concentration of the competitors (TDDT). It is seen that even at very low concentration of competitors (1 % of AT) supralinearty is observed. If the competitors are eliminated by pre-filling the TDDTs by a high pre-dose we would have $m = M$ in (6.6). This would lead to

$$I = n = N[1 - \exp(-\alpha D)]$$

Curve 5 in Fig. 6.4 which shows a linear growth till approaching saturation, is the plot of this equation. This type of behavior (superlinear growth in well annealed virgin phosphor and linear growth in the pre-irradiated and partially read out samples) is seen in many TL phosphors. LiF:Mg, Ti(TLD-100) is an outstanding example of such behavior.

6.5 Combination of Linear and Superlinear Growth of TL Intensity

In the preceding section both the traps and the RC were assumed to be isotropically distributed together with the competitor traps or the nonluminescent RC. The existence of competitor traps or nonluminescent RC in such a distribution, leads to superlinear growth of TL intensity throughout the sub-saturation part of the growth curve. However, when the competitors are eliminated, the growth curve becomes linear (see curve 5 in Fig. 6.4). We may therefore safely imagine that a system consisting of two sets of traps—one which face no competition and other facing competition, but both having the same E value, would have two components in its growth curve—one linear and the other superlinear.

A theoretical model for linear–superlinear growth of TL intensity based on the local and nonlocal recombination was first developed by Mische and McKeever [7]. Their model proceeds as follows: Consider that the traps which form a certain fraction J of their total population n are each at a distance less than s from an RC. The value of s is small enough (they take $s = 10^{-8}$ m in a numerically solved case) so that no other defect of any type—trap, RC, or competitor exists within the volume $\frac{4}{3}\pi s^3$. Further, it is assumed that s is not so small that it could disturb the potential around the trap, so that the E value of all the traps numbering n remains the same. The probability that the released carrier from the volume $\frac{4}{3}\pi s^3$ will recombine with its paired RC within this volume is given by ${}^{\sigma_h}/_{4\pi s^2}$, where σ_h is the capture cross section of the RC. The probability that the released carrier escapes from this volume and recombines elsewhere in the sample volume is

$$\left(1 - \frac{\sigma_h}{4\pi s^2}\right)\mu\sigma_h h \tag{6.10}$$

where h is the concentration of the RC and μ is the mean free path of the released carrier between the capture events (capture by RC or by a competitor). It is given by

$$\mu = \frac{1}{(\sigma_h h + \sigma_2 n_2)} \tag{6.11}$$

where n_2 is the concentration of the competitors and σ_2 is their capture cross section. The total probability for undergoing recombination for the released charge carrier from the above defined volume including both with the paired RC within this volume and elsewhere in the sample therefore is

$$P_j = Jn\left[\frac{\sigma_h}{4\pi s^2} + \left(1 - \frac{\sigma_h}{4\pi s^2}\right)\mu\sigma_h h\right] \tag{6.12}$$

The probability for recombination of the remaining $(1 - J)$ fraction of the electron population n, then is

$$P_{(1-j)} = n(1 - J)m\sigma_h h \tag{6.13}$$

The sum of (6.12) and (6.13) gives the total probability, S of luminescence producing recombinations and is given by

$$S = n\left[J\frac{\sigma_h}{4\pi s^2} + \left(1 - J\frac{\sigma_h}{4\pi s^2}\right)\mu\sigma_h h\right] \tag{6.14}$$

If we have $h\sigma_h \gg \sigma_h n_2$, the value of μ would reduce to $1/h\sigma_h$ (see (6.11)) and we get S = n. This means we would have TL intensity growing linearly with dose if the traps fill linearly with dose. On the other hand, when competitors are far in excess of RC ($\sigma_2\, n_2 \gg \sigma_h\, h$), one would have $\mu = \dfrac{1}{\sigma_2 n_2}$ and (6.14) becomes

$$S = n\left[J\frac{\sigma_h}{4\pi s^2} + \left(1 - J\frac{\sigma_h}{4\pi s^2}\right)\frac{\sigma_h h}{\sigma_2 n_2}\right] \tag{6.15}$$

This equation consists of the sum of a linear term $\left(nJ\sigma_h/(4\pi s^2)\right)$ and a nonlinear term $n\left[\left(1 - J\frac{\sigma_h}{4\pi s^2}\right)\frac{\sigma_h h}{\sigma_2 n_2}\right]$. As the concentration n of the AT increases with dose and that of the competitors n_2 decreases due to getting filled up, the nonlinear term would begin to dominate. The authors [7] assume that the traps and the RC get filled up with dose in a way similar to (6.4) and (6.5). Further when the competitors are assumed to be traps, they too get filled up with dose in the same manner. If the total concentration of the competitor traps is taken as N_2, and their filled concentration n'_2, one would have the concentration of the competitors $n_2 = N_2 - n'_2$. Using selected sets of parameter values in (6.15), Mische and McKeever [7] have shown the TL growth curves which are linear at low doses and become superlinear as the dose is increased. The curves become sublinear before reaching saturation. It may be noted that the designations of the parameters are changed in order to maintain the uniformity in this book.

6.6 Interactive and Noninteractive Traps and Their Role in Supralinearity

In order to obtain linear-superlinear-sublinear-saturation type of TL intensity growth termed as supralinear growth, Sunta et al. [25], assume two sets of AT, both having the same activation energy E and the same filling rate constant α. From one set the released carriers recombine locally with the spatially associated RC. This set is called noninteractive. From the other the released carriers are delocalized. In the delocalized domain the carriers may either recombine with RC to emit luminescence or be captured by the TDDT nonradiatively. The TDDTs thus act as competitors. This set is termed as interactive. The model with the two sets of AT together with the TDDT may be called the partly interactive multi trap system (PMTS). The difference between the model of Sunta et al. [25] and that of Mische and McKeever [7] is that the latter authors assume only one type of traps, some of the released carriers from which recombine locally to emit luminescence. The remaining get delocalized, out of which some recombine to emit luminescence and others are captured nonradiatively by the competitors. These competitors are undefined. In contrast those in the model of Sunta et al. are physically defined TDDT. Secondly, Mische and McKeever assume the concentration of the competitors far in excess of that of the RC. In the PMTS model of Sunta et al. there is no restriction on the relative concentrations of the competitors and RC. Apart from explaining the supralinearity in the growth of TL intensity, a special advantage of this model is that, it is able to derive the theoretical expressions for the supralinearity and the predose sensitization factors. In this model the AT, both of noninteractive and the interactive sets, are assumed to have the same activation energy and the same filling rate constant α. The activation energy of the TDDT as well as its filling rate constant ε may be different from those of the AT. The physical model proposed is illustrated in Fig. 6.5. The total concentration of the interactive traps is designated as N_i and that of noninteractive traps as N_n. The corresponding filled concentrations at any dose are n_i and n_n. Similarly the total and the filled concentrations of TDDTs are M and m. According to the charge neutrality condition, the concentration of RCs for interactive traps is $(n_i + m)$ and that for noninteractive traps it is n_n. The photon energy of the light emitted from both these RCs is implicitly the same. The transition coefficients designated as A_h and A_m are for recombination with RC and recapture in TDDT, respectively, (Sunta et al. use γ and ε in place of A_h and A_m , respectively, in their papers [25, 26]. The present notations are used for the sake of uniformity in this book). The TL intensity contribution of interactive traps is already discussed in Sect. 6.4. Let it be designated as I_i. The value of I_i may be obtained by substituting the expressions for n and m from (6.4) and (6.5), respectively, into (6.6). The contribution I_n of the noninteractive traps at any αD is given by (6.4). It may be noted that we use n_n and N_n for noninteractive traps and n_i and N_i for interactive traps in (6.4). Using $N_i = KN_n$ and $M = aN_i = aKN_n$ and adding the contributions I_i and I_n from interactive and noninteractive traps, expression of the total intensity I_t (area of the glow peak) turns out to be

$$I_t = I_n + I_i = N_n\left(1 - e^{-\alpha D}\right)\left[1 + K\frac{(1 - e^{-\alpha D}) + a(1 - e^{-\varepsilon D})}{(1 - e^{-\alpha D}) + a(1 - e^{-\varepsilon D}) + a\left(\frac{A_m}{A_h}\right)e^{-\alpha D}}\right] \tag{6.16}$$

Sunta et al. [25] examined the effect of a and K factors on the profiles of I_t as a function of dose D. Assuming $\varepsilon = \alpha$ and $A_m = A_h$ (6.16) gets simplified to

$$I_t = N_n\left(1 - e^{-\alpha D}\right)\left[1 + K\frac{(a + 1)(1 - e^{-\alpha D})}{a + 1 - e^{-\alpha D}}\right] \tag{6.17}$$

This simplification does not affect the profiles of the changes in the TL intensity growth curves I_t, caused by different values of K and a. The I_t versus dose profiles for different values of a and K are given in Figs. 6.6a, b. In calculating these

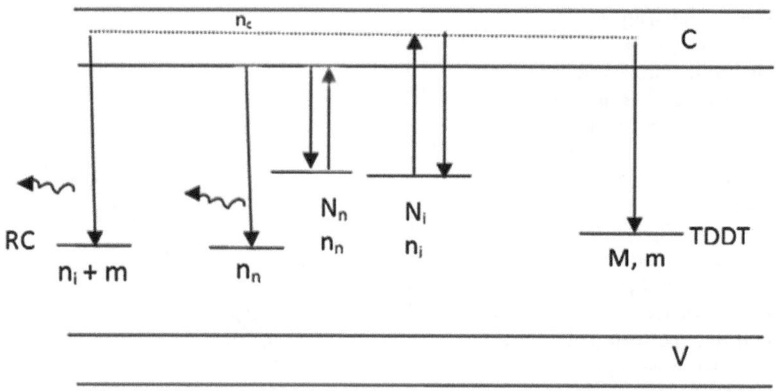

Fig. 6.5 Energy level scheme showing the interactive and the noninteractive traps. The total populations of these are shown, respectively, as N_i and N_n, their filled concentrations, respectively, are n_i and n_n. The total concentration of TDDTs is shown as M and their filled concentration as m. The *wavy arrows* show the TL emission

growth profiles the parameter α is taken as 10^{-3} Gy^{-1}, so that the glow peaks saturate around 10^4 Gy. It is noteworthy that various shapes of the growth profiles are possible, depending on the values of a and K. In Fig. 6.6a which is for $a = 10$ and different values of K, it is seen that supralinearity decreases with decrease in K which means decrease in the concentration of the interactive traps. In Fig. 6.6b, it is seen that at a given value of K, as a decreases TL intensity is increased, and the superlinear part begins progressively at lower and lower doses (see curves 3, 4 and 5). This is because when there are smaller number of competitors (low values of a) the TL yield of the interactive traps which produce superlinearity, would increase even at low doses. It is seen that as long as we have $K > 1$, a supralinear pattern of growth would be observed even if a is very small (but not 0). Whereas the model of Kristianpoller et al. [24] requires the population of the competitors to be much greater than that of the AT, in the present model there is no restriction on the values of these parameters.

To sum up we see that the shape of the TL intensity growth curve is dependent on the factors a and K where, as assumed earlier, $a = M/N_i$ and $K = N_i/N_n$. When either a or K is zero, the supralinearity vanishes. Stated in words this means supralinearity vanishes when either TDDT or interactive traps are absent. Greater are the values of a and K factors, greater is the supralinearity (see Fig. 6.6a). The maximum value of supralinarity factor however, depends largely on K. The saturation level of the TL intensity growth curve also depends on the value of K. Interestingly for a given value of K, smaller is the value of a earlier does the supralinearity begin (see Fig. 6.6b). We thus see that with this model one can get various shapes of intensity growth curves depending on the values of a and

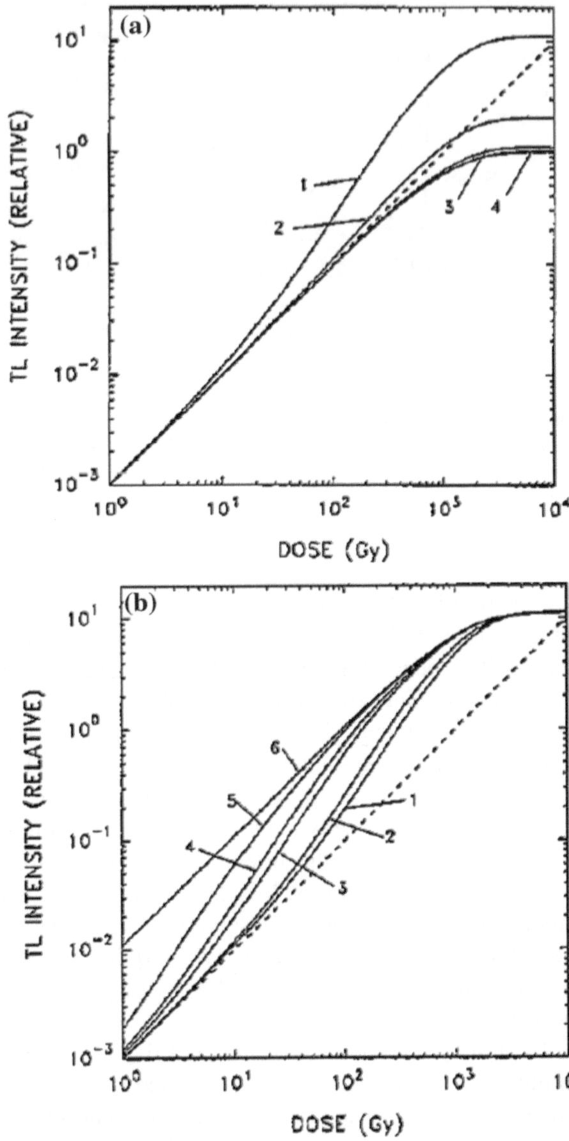

Fig. 6.6 **a** TL intensity I_t versus dose for $a(= M/N) = 1$ and different $K(= N_i/N_n)$ values. *Curves 1–4* are for $K = 10$, 1, 0.1 and 0.01 to 0, respectively. **b** TL intensity I_t versus dose for $K = 10$ and different values of a. *Curves 1–6* are for $a = 10$, 1, 0.1, 0.05, 0.01 and 0, respectively. The *broken lines* in (**a**) and (**b**) show the extended line of linearity. It may be seen that as long K is large enough supralinearity would be observed even if the value of a is small (but not 0). *Curves 3* and *4* in (**a**) do not show supralinearity, because in this case the interactive traps are nearly or totally absent ($K \approx 0$). The TL response in this case is only from the noninteractive part of the active traps which face no competition from the TDDT [25, Fig. 3a, b]. © IOP publishing Ltd. Reproduced by permission. All rights reserved

K. When K approaches zero the saturation level of the growth curve is determined by the value of N_n alone. In curve 1 in Fig. 6.6a and curves 1 and 2 in Fig. 6.6b superlinear part begins at a dose of about 5 Gy. It means that a TL phosphor with $\alpha = 10^{-3}$ Gy^{-1} and with the intensity growth curve ranging from 1 mGy to about 10^3 Gy, will have about three and half decades of linear range before superlinearity begins at about 5 Gy of dose (see Fig. 6.6a, b).

6.7 Supralinearity and Sensitization

Increase in sensitivity of the glow peaks of phosphors by pre-irradiating and partial annealing (heating up to a temperature sufficient to erase the glow peak under study and leaving the higher temperature peaks as residual) is well known in the case of many phosphors (e.g. LiF:Mg, Ti, CaF_2:natural (fluorite), CaF_2:Dy, $CaSO_4$:Dy, Mg_2SiO_4:Tb etc.) [5, 7, 8, 31–33]. The supralinearity discussed in the preceding section is caused particularly by filling of the interactive part, N_i of the AT and the traps of the TDDT, M. The filling of both these causes a quadratic increase in the TL yield per unit dose (response factor). Filling of TDDTs reduces the number of competitors and thus helps in increasing the TL yield. At low doses the competition being high, the contribution of the interactive part of the AT n_i becomes an under-response because of the loss of the released carriers due to capture by the empty TDDTs. A progressive reduction in the under response by the filling of the TDDTs during the irradiation, thus causes the so-called sensitization. We see that this process of sensitization progresses continuously during irradiation in an interactive trap system involving TDDTs. During the readout heating this leads to the TL intensity per unit dose to increase at a rate greater than linear until the ATs begin to saturate. The greater than linear growth is described as superlinearity. In the total intensity growth curve (produced both by the noninteractive and interactive traps), superlinearity begins at a dose at which the intensity contribution of the interactive traps n_i nearly overtakes that from the noninteractive traps n_n. This can, however happen only when $N_i > N_n$, where N_i and N_n are the total concentrations of the interactive and the noninteractive type of the AT, respectively. In the mathematical treatment above this means $K > 1$. The total intensity growth curves for different K and a values are shown in Fig. 6.6.

6.7.1 Supralinearity Factor

As described in the previous section when the contribution of the noninteractive part of the active trap population n_n is added to that of the interactive part n_i, the total TL intensity I_t (D) at a given dose D is given by (6.16) or its simplified form (6.17). The quantity I_t $(D)/D$, is called the relative response factor [3] or the SF. The theoretical equation for SF may be written as

$$SF = \frac{I_t(D)}{N_n \alpha D} = \frac{(1 - e^{-\alpha D})}{D}\left[1 + K\frac{(a+1)(1 - e^{-\alpha D})}{a + 1 - e^{-\alpha D}}\right] \tag{6.18}$$

where $N_n\alpha D$ in the denominator represents the straight line drawn by extending the linear part of the TL intensity growth curve. One may simply use D in place of $N\alpha D$, in the denominator, since N and α are constants. The values of SF using this equation are shown in Fig. 6.7 for $K = 10$ and $a = 10$, 1 and 0.1 and also for $K = 1$ and $a = 1$.

6.7.2 Pre-dose Sensitization Factor

In pre-dose sensitization measurement, the pre-irradiation and the subsequent heating bleaches out the glow peak for which the pre-dose sensitization is to be studied and leaves the filled TDDTs, m as residual, thus reducing the concentration ($=M - m$) of the competitors. The reduction depends on the pre-dose given which determines the value of filled TDDTs ($=m$). As a result of this reduction in competitors, there is increase in radiative recombination during heating of the test dosed sample leading to an increase in its TL yield. The TL yield from a pre-dosed sample produced by a test dose, when the AT of the said sample belong to the interactive system may be derived from (6.6). In this equation take n as the concentration of the AT filled by the test dose and m as that of the TDDTs filled by a pre-dose D. The test dose is required to be much smaller than the pre-dose so that the effect of the test dose on the sensitization caused by the pre-dose is negligible. For simplicity let us assume $A_m = A_h$ in (6.6). The small value of test dose renders $n \ll M$. The test dose being the same for the samples exposed to different values of pre-dose, the value of n becomes constant. If we designate this constant value of n by c and use $A_m = A_n$ and $c \ll M$, (6.6) becomes

$$I_i = c(c + m)/M \tag{6.19}$$

where I_i is the TL intensity produced by the interactive part of the AT and m is the concentration of the deep traps filled by the pre-dose and is the only variable in this equation. The value of m as a function of pre-dose D is given by (6.5), thus (6.19) becomes

$$I_i = \left(c^2/M\right) + c[1 - \exp(-\varepsilon D)] \tag{6.19a}$$

Since we are dealing with a system in which both interactive and noninteractive AT are involved, one needs to add to (6.19a), the contribution c_n of the noninteractive part produced by test dose to get the total intensity I_t. In (6.19a) we may use $c = K\,c_n$ (refer Sect. 6.6 for the meaning of K), we may then express the total intensity I_t, as

$$I_t = c_n + \frac{(Kc_n)^2}{M} + Kc_n(1 - e^{-\varepsilon D}) \tag{6.20}$$

In this equation second term is negligible (since $c_n \leq 10^{-3}\,N_n$ and $M = aKN_n$ and $a \geq 1$), (6.20) may therefore be simplified as

$$I_t = c_n\left[1 + K\left(1 - e^{-\varepsilon D}\right)\right] \tag{6.21}$$

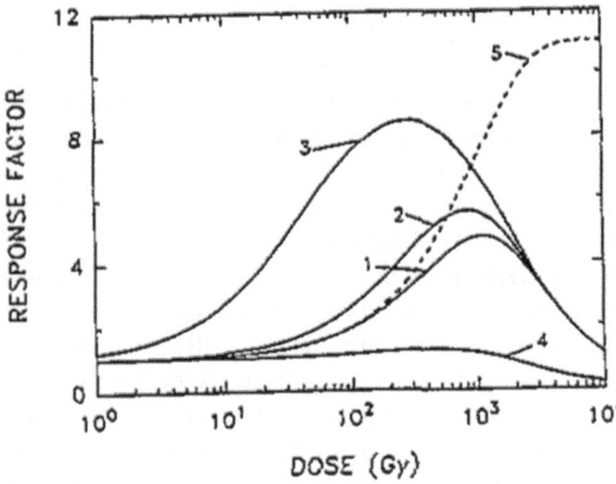

Fig. 6.7 Effects of *K* and *a* on supralinearity factor (*SF*). *Curves 1–3* are for $K = 10$ and $a = 10$, 1 and 0.1, respectively, *curves 4* is for $K = 1$ and $a = 1$. *Curve 5* gives the pre-dose sensitization factor (*PDSF*) as a function of pre-dose when $K = 10$. The *X*-axis for *PDSF* curve is to be read as pre-dose (Gy). These *curves* show that the value of *SF* increases with the decrease in *a*. At the same time its peak value shifts progressively to lower temperatures. The decrease in *a* means decrease in concentration of TDDT, which act as competitors [25, Fig. 4].

The quantity I_t/c_n provides a theoretical expression for the pre-dose sensitization factor (*PDSF*) as a function of pre-dose *D* viz.

$$PDSF = 1 + K\left(1 - e^{-\varepsilon D}\right) \tag{6.21a}$$

The experimentalists represent the *PDSF* by S/S_0 where *S* is the TL intensity produced by a given test dose to a sample which has been pre-irradiated by dose *D* and S_0 is the TL intensity produced by the same test dose when the sample is not given pre-dose. Equation (6.21a) shows that the sensitization effect depends primarily on the factor K. A plot of (6.21a) is shown in Fig. 6.7 as curve 5 for $K = 10$ and $\varepsilon = 10^{-3}\ \mathrm{Gy}^{-1}$. The maximum value of *PDSF* from (6.21a) is $(1 + K)$ and it is this value which we see at the peak of the theoretical *PDSF* profile. The experimental curve S/S_0 versus pre-dose for real phosphors usually shows a decline after reaching the maximum due to the radiation damage to the phosphor at high pre-doses which leads to reduction in TL yield. The extent of damage effect depends on the phosphor material. It is not accounted in the theoretical curves of Fig. 6.7. Note that maximum of the *PDSF* is shifted to the higher dose side of the maximum of the SF. The result is commensurate with the experimental observations in LiF:Mg, Ti [30, 34]. This result provides a theoretical explanation for the observed difference in the position of the maxima in the curves of the *SF* and the *PDSF*.

6.7.3 Sensitization Factor (S_nF)

As discussed in the beginning of Sect. 6.7, sensitization takes place progressively with increasing dose. Increase in dose fills the AT as well as the TDDTs. The filling of the latter leads to a progressive reduction in the number of vacant TDDTs (=$M - m$) which act as competitors. The combined effect causes the TL intensity to rise more than linearly with dose and this is what we call sensitization. A new factor called the sensitization factor (S_nF), distinct from *SF* and *PDSF*, is introduced here which provides a unified explanation for the similarities and dissimilarities in the characteristics of the factors called the *SF* and the pre-dose sensitization factor (*PDSF*). First let us understand the difference between S_nF and the *SF*. This may be understood with the help of the illustration given in Fig. 6.8a in which the typical profiles of I_n, I_l and I_t are given as a function of dose *D*, where I_t is the total TL intensity (which means intensity contributions both from interactive and noninteractive traps) produced in a sample at any given dose *D* (see (6.16) and (6.17)) and I_n is the intensity contribution of the noninteractive traps n_n alone. Curve I_l is the line of linearity extended from the initial linear coincident part of the growth curves I_t and I_n. In the model used here (refer Sect. 6.6), curve I_n, which is produced by noninteractive traps n_n rises linearly until the saturation effect begins. On the other hand the intensity I_i produced by the interactive traps n_i rises superlinearly. The curve I_i lies under the curve of I_n at low doses. The total TL intensity I_t is the sum of these. In recording the TL intensity growth of an experimental sample showing the supralinear characteristics, we can see only the profile of I_t. Let us assume that the total concentration N_i of the interactive traps is equal to 10 N_n. This means that the curve I_t would saturate at about 10 times higher intensity than the curve I_n. We may now give the values of *SF*(*D*) and $S_nF(D)$ at any given dose *D* using the three curves of Fig. 6.8a viz.

$$SF(D) = I_t(D)/I_l(D) \tag{6.22a}$$

and

$$S_nF(D) = I_t(D)/I_n(D) \tag{6.22b}$$

The theoretical expression for *SF*(*D*) is already given in (6.18). Similarly, using the expressions of I_t and I_n as given by (6.16) and (6.4), respectively, the theoretical expression for $S_nF(D)$ turns out as

$$S_nF(D) = 1 + K\left[\frac{(\mathrm{a}+1)(1-\mathrm{e}^{-\varepsilon \mathrm{D}})}{\mathrm{a}+1-e^{-\alpha D}}\right] \tag{6.23}$$

We may also manually produce the curves for *SF* and S_nF using the three curves I_t, I_n and I_l of Fig. 6.8a and (6.22a) and (6.22b). The curves so produced are

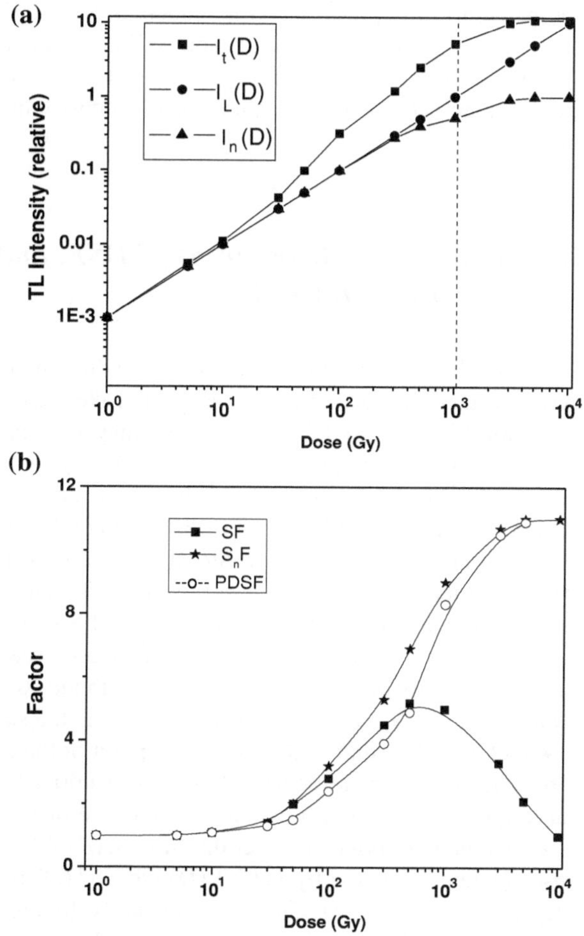

Fig. 6.8 **a** Illustration for deriving S_nF and SF. Curve $I_t(D)$ is the total intensity. The straight line $I_l(D)$ is the extension of the initial linear part of the curves $I_t(D)$ and $I_n(D)$. Curve $I_n(D)$ is the intensity growth profile for the noninteractive part of the active traps n_n. The curves $I_n(D)$ and $I_t(D)$ are drawn by using $\alpha = 10^{-3}$and $\varepsilon = 10^{-4}$. **b** Theoretical curves of SF and, S_nF as a function of dose D. Values of $PDSF$ as a function of pre-dose D also are plotted for comparison. For PDSF X-axis should be read as Predose (Author's unpublished work)

given in Fig. 6.8b. If we compare the equation of SF (see (6.18)) with that of S_nF, the former becomes identical to that of the latter at low values of αD at which $(1 - e^{-\alpha D})$ outside the square brackets in (6.18) becomes equal to αD. Thus at low doses S_nF and SF are coincident, which we see practically in Fig. 6.8b. At higher doses, as seen from the divergence between the profiles of I_t and I_n, (see Fig. 6.8a) the value of S_nF increases rapidly with dose, whereas that of SF begins to fall after reaching a peak. The fall SF appears because the straight line I_l continues to rise indefinitely, even when I_t begins to saturate. As a result the maximum of SF appears on a much lower dose than that of S_nF. In Fig. 6.8b are plotted also the values of $PDSF$, in which the pre-doses are the same as the dose values for which S_nF is calculated as described above. We see that the curve of S_nF lies slightly above that of $PDSF$. This is because the 'test dose' for calculating the S_nF values is as high as the 'pre-dose', whereas the test dose in the case of $PDSF$ is negligible in comparison to the pre-dose. This renders $S_nF > PDSF$. At very low doses all the three curves, namely *SF*, *PDSF* and S_nF are coincident as may be obvious also from the growth curves given in Fig. 6.8a. The maximum values of S_nF and *PDSF* are the same because of the saturation of intensity in the growth profiles of both I_t and I_n.

6.7.4 Comparison Between SF, PDSF, and S_nF in a Real Material

So far we have discussed the effects of radiation sensitization theoretically. We may now take the case of a real material. Take the example of the well-known TL phosphor LiF:Mg, Ti (TLD-100). We may compare in this material the three factors, namely the *SF*, the S_nF and the *PDSF*. The results may be discussed in the perspective of the theoretical results given above. As is well known the two factors, namely the *SF* and the *PDSF* may be experimentally measured. As explained above, to find the profile of S_nF we need the curve for I_n in addition to the experimentally measurable curve of I_t. The curve for I_n is essentially a theoretical curve given by (6.4), which needs the values of N_n and the growth rate constant α (see (6.4)). As discussed later in Sect. 6.8.2, the value of α for LiF:Mg, Ti (TLD) phosphor has been found to be about 10^{-3} per Gy. In Sect. 6.8.1 it is found that $K = N_i/N_n = 13$, which means that the saturation point for the curve of I_n would lie below that of the saturation point of the experimental TL growth curve I_t by a factor of 14 (refer Sect. 6.8.1). The initial linear part of the experimental curve I_t (which essentially is part of the curve of I_n) of the fresh phosphor sample may then be extrapolated to get the full curve of I_n. The values of S_nF at different doses *D* then are given by $I_t(D)/I_n(D)$ where $I_t(D)$ is the experimentally measured TL intensity at dose *D*. The experimentally found values of *SF* and *PDSF* and those of S_nF at different doses are plotted in Fig. 6.9 for peak 5 (appearing at about 200 °C) of the LiF:Mg, Ti (TLD-100) phosphor. In the case of *PDSF* the *X*-axis should be read as pre-dose and *Y*-axis as *PDSF* for curve 3 of Fig. 6.9. Theoretically the *PDSF* and the S_nF values on saturation, approach a plateau as seen in Fig. 6.8b. However in the case of the phosphor LiF:Mg, Ti (TLD-100), the *PDSF* as well as S_nF curves fall after reaching their maxima. This fall is due to the radiation damage to the phosphor at high pre-doses which cause decrease in the value of TL intensity. The radiation damage causes decrease in the values of I_t at high doses. This leads the values of S_nF to fall. If radiation damage is assumed to be the same also in I_n, the curve of S_nF would not show a fall. In the intermediate dose range, as stated before, the values of S_nF are higher than those of *PDSF*, because the 'test dose' in the case of S_nF is about the same as the 'pre-dose' (both being comparatively high), whereas in the case of *PDSF* the test dose is negligibly small.

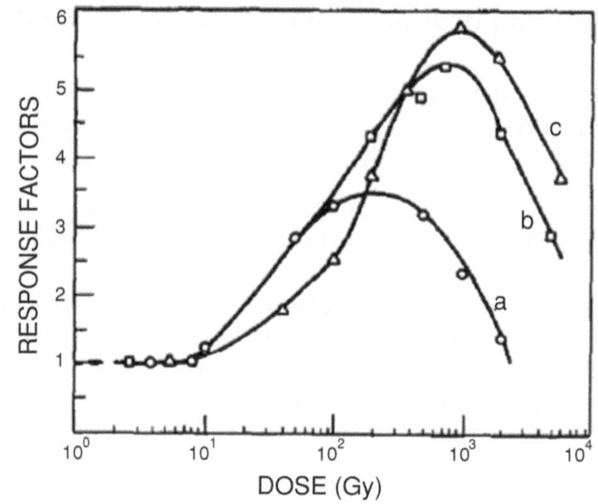

Fig. 6.9 Observed profiles of the three factors: **a** *SF*, **b** S_nF and **c** *PDSF* for peak 5 of LiF: Mg, Ti (TLD-100). *Y*-axis is marked as response factor for all the three factors. *X*-axis gives the dose given. For PDSF it means the pre-dose [30, Fig. 7]. ©Elsevier science Ltd. Reproduced by permission

Although a mathematical relation between *SF* and *PDSF* had not been obtained by the earlier workers, a close relation between the two had been well recognized in the early stage of TL dosimetry development using LiF. Suntharalingam and Cameron [16] had found that both *SF* and *PDSF* are inversely related to LET (linear energy transfer) of the radiation used in inducing the TL. This similarity in the behavior of these two factors led them to suggest a close relationship between them. The relation between them may be inferred also from the fact that a sample exposed to saturation pre-dose to cause maximum sensitization shows neither supralinearity [8] nor further pre-dose sensitization. Notwithstanding these facts Lakshmanan et al. [34] and Niewiadomski [35] have expressed that the mechanism involved in SF and PDSF may be altogether different for each of these two phenomena. The reasons for their reservations are cited to be the difference in the values of *SF* and *PDSF* as well as the dose values at which the maxima in these two factors are reached. As shown in Figs. 6.7 and 6.8b, these differences in their behavior are expected from the theory itself. A direct relation between the supralinearity and the pre-dose sensitization is further clarified by the new factor called the S_nF. This factor as well as the *PDSF* are both caused by the radiation sensitization. Their profiles are nearly the same. The maxima of both these factors have nearly the same magnitude and lie at the same dose (see Fig. 6.8b). Both are displaced to the higher dose side of the *SF*. The reason why the peak value of S_nF and dose at which the peak is reached, is greater than that of *SF* is already explained above. The same explanation may apply to the difference between *SF* and *PDSF*.

6.8 Application of the Model to the Real Samples

The model consisting partly of interactive and partly of noninteractive traps described in the preceding pages may be applied specifically to such materials, whose glow peaks exhibit the following characteristics:

(1) The TL response of freshly annealed sample is linear at lower doses and is followed by superlinear, sublinear and saturation regions as the dose is increased.
(2) Pre-exposure to radiation followed by partial annealing to erase the lower temperature TL peaks sensitizes the lower temperature peaks. Sensitization increases as the pre-dose increases till reaching saturation.
(3) Supralinearity of the lower temperature peak is eliminated when the sample is sensitized to its maximum by irradiation with the saturation pre-dose.

These characteristics are seen in many TL phosphors exposed to low LET (linear energy transfer) radiations such as X and gamma rays of photon energy E > 500 keV. Well known examples of these phosphors are LiF:Mg, Ti [8, 31], $CaSO_4$:Dy, $CaSO_4$:Tm, CaF_2:Dy [32, 33], Mg_2SiO_4:Tb [5] and fluorite mineral [4]. Sunta and co-workers [31] have applied this model to the well known TLD phosphor LiF:Mg, Ti to determine the parameters α, ε (the trap filling rate constants), K (ratio between the concentration of the interactive type of AT N_i and that of noninteractive type N_n) and a (ratio between the concentration of the TDDTs and that of interactive type of AT N_i). These parameters are found for peak 5 ($T_m \approx 200$ °C) of this phosphor. They have applied this model also to the mineral CaF_2 [36] TLD phosphor. Application to LiF:Mg, Ti is described briefly in the following. It provides the methodology which may be applied also to other phosphors having the above stated characteristics. Materials like quartz, however, may not come within the purview of this model, since the effect of irradiation after a high temperature annealing in this material [10] are quite different from that described in above mentioned materials.

6.8.1 Ratio Between the Concentrations of Interactive Traps N_i and Noninteractive Traps N_n (N_i/N_n Represented as Factor K)

The factor K was introduced in Sect. 6.6 according to which $K = N_i/N_n$, where N_i and N_n are the total concentrations of interactive and noninteractive parts, respectively, of the active trap population. The value of K may be estimated from the TL intensities produced by the filled concentrations (designated by n_i and n_n respectively) of these parts of traps individually. Figure 6.10 shows the TL intensity growth curves of the so-called peak 5 ($T_m \approx 200$ °C depending on the heating rate) of LiF:Mg, Ti. Curve 1 is for well annealed (400 °C, 1 h) phosphor and Curve 2 for the pre-dose sensitized (with saturation pre-dose of 10^3 Gy) sample. As may be seen both the curves reach their maximum intensity level at about the same dose and decline thereafter due to the damage effect of high dose. Damage to TL by a dose of 10^3 Gy is about 50 % [31]. Curves 3 and 4 are the damage corrected versions of curves 1 and 2, respectively. Comparison of TL intensities of curves 1 and 4 at the dose level of less than 3 Gy (linear region of curve 1) reveals that the actual sensitization by the pre-dose is by a factor of 14. Without the damage correction this factor would be 6.5 (compare curve 2 with curve 1 at dose <3 Gy). The S_nF 14 may be used to arrive at the relative concentration factor *K* between the interactive and noninteractive parts of the active trap population. The intensity values on curves 4 consist of contributions both from noninteractive and interactive parts of the active trap population. These intensities may be designated by I_n and I_i , respectively. In this case the contribution I_i of the interactive part is after elimination of competition by the vacant TDDT, since the TDDT are pre-filled by the saturation pre-dose. Elimination of competition means that all the traps consisting of population n_i produce luminescent recombination. The intensity values on curve 1 at dose <3 Gy may be considered to be due only to noninteractive trap population. The contribution of the interactive part of AT on curve 1 at these low dose levels is negligible due to the competition by the vacancies in the TDDT. If we designate the intensity values on curve 4 as I_4 and use the values of I_4, I_i and I_n at dose values less than 3 Gy, we may write

$$I_i + I_n = I_4 \tag{6.24}$$

As stated above we have

$$I_4/I_n = 14 \tag{6.25}$$

From these two relations, we have

$$I_i/I_n = 13 \tag{6.26}$$

I_n and I_s are directly proportional to n_n and n_i, since the trap filling rate constants α is taken the same for the noninteractive and interactive type of traps. As stated before these traps are similar in character, except that the former are spatially close to the RC and latter not. The values of n_n and n_i are directly proportional to N_n and N_i , respectively. It therefore means that $N_i/N_n = 13$, which ratio has been designated as K in the preceding.

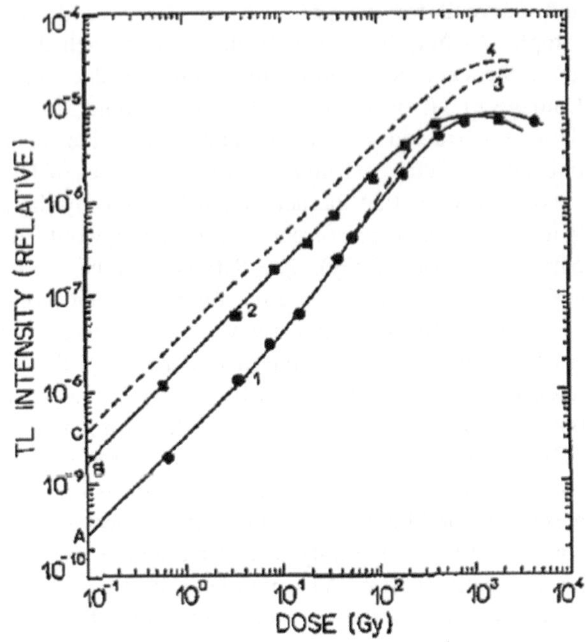

Fig. 6.10 TL intensity as a function of dose for glow peak 5 of LiF:Mg, Ti (TLD-100). *Curve 1*: Sample annealed at 400 °C for 1 h and 100 °C for 2 h, *curve 2*: sample of curve 1 irradiated to 10^3 Gy and then heated to bleach glow peaks up to 350 °C, *curves 3* and *4* are obtained from *curves 1* and 2, respectively, after correcting these curves for radiation damage [31, Fig. 2].

6.8.2 Trap Filling Rate Constants α and ε

The trap filling rate constant α of the traps belonging to a given glow peak may be determined either from the TL intensity growth profile of the said glow peak or from the OA growth profile of the trap, if the OA band is identifiable for the given trap. The TL growth profile in the case of peak 5 of LiF:Mg, Ti is curve 4 of Fig. 6.10, which gives the damage corrected intensity of a pre-dosed (with saturation pre-dose) sample. Fitting (6.4) to this curve yields the α value equal to $(1.35 \pm 0.12) \times 10^{-3}$ Gy^{-1}. Data taken from the TL growth profile measured by Lakshmanan et al. [37] for a similar sample gave a value of $(1.19 \pm 0.04) \times 10^{-3}$ Gy^{-1}. In LiF:Mg, Ti, peak 5 has been identified to be due to the defect center which shows an OA band at 310 nm [38–40]. The OA growth data for this band produced a value of $(0.93 \pm 0.04) \times 10^{-3}$ Gy^{-1}. This was, however, by combining the data of different dose ranges from two different publications [40, 41]. It may therefore be subject to error. The average of these three values is 1.16×10^{-3} Gy^{-1} which may be considered as reasonable.

For determining the filling rate constant ε for the TDDT, (traps of 400 °C glow peak of LiF:Mg, Ti) extensive data similar to those for peak 5 are not available. It is, however seen that the 400 °C peak saturates at an order of magnitude higher dose than peak 5 [8, 42]. It implies that the ε value should be an order of magnitude lower than α (see (6.4) and (6.5)). This conclusion is supported by the OA data [40, 42] which show that Z_3 band (225 nm) in LiF Mg, Ti which appears to be associated with the 400 °C TL peak, saturates at a higher dose than 310 nm OA band. The latter band as said before belongs to the defect center (trap) of peak 5.

6.9 Ratio Between the Concentration M of TDDT and Concentration N_n of the Interactive Part of the Active Traps Represented by a, ($M/N_i = a$)

The constant a has been introduced in Sect. 6.6 according to which a $= M/N_i$. The value of a may be found by using the ratio R between the TL intensity I_i obtained in the presence of competition by TDDT and the intensity I_s when the competition is eliminated by fully pre-filling the TDDT by saturation pre-dose. Both these intensities I_i and I_s, pertain to the interactive part of the active trap population which is designated by n_i (see Sect. 6.6). As described earlier the value of I_i is given by (6.6). When the TDDTs are fully prefilled by pre-dose, one would have $m = M$ in (6.6) and therefore we may take $I_s \equiv n_i$, the ratio R between the two intensities, is then given by

$$R = \frac{I_i}{I_s} = \frac{(n_i + m)A_h}{(M - m)A_m + (n_i + m)A_h} \tag{6.27}$$

Substituting the equations for n_i and m from (6.4) and (6.5), respectively, (use n_i and N_i in place of n and N in (6.4)) and using $M = \mathrm{a}N_i$ we obtain

$$\mathrm{R} = \left[\frac{(1-e^{-\alpha D}) + \mathrm{a}(1 - e^{-\varepsilon D})}{(1-e^{-\alpha D}) + \mathrm{a}(1 - e^{-\varepsilon D}) + \mathrm{a}\frac{A_m}{A_h}e^{-\varepsilon D}}\right] \tag{6.28}$$

Using αD and $\varepsilon D \ll 1$ in (6.28) leads to

$$\mathrm{R} = \frac{\alpha\mathrm{D}\,(\,1 + \mathrm{a}\frac{\varepsilon}{\alpha})}{\alpha\mathrm{D}\,(\,1 + \mathrm{a}\frac{\varepsilon}{\alpha}) + a\frac{\mathrm{A_m}}{\mathrm{A_h}}}) \tag{6.29}$$

To determine the value of R for the supralinear growth curve of LiF:Mg, Ti (TLD-100) which is given as curve 3 after correction for radiation damage in Fig. 6.10, it is required to split this curve into two components namely, (1) $I_n$versus dose for noninteractive part of AT and (2) I_i versus dose for interactive part of AT. Full curve of component No. 1 is found by extrapolating the initial linear part of curve 1, which is given by $N\alpha D$, by using $\alpha = 1.16 \times 10^{-3}$ Gy^{-1} (refer Sect. 6.8.2). Component No. 2 is found by subtracting component 1 from curve 1. The two components are shown by curves 3 and 4, respectively, in Fig. 6.11. Curve 2 in this figure is the same as curve 4 of Fig. 6.10. It merges with curve 5 which is for I_s. As explained in Sect. 6.8.1 the intensity values in curve 5

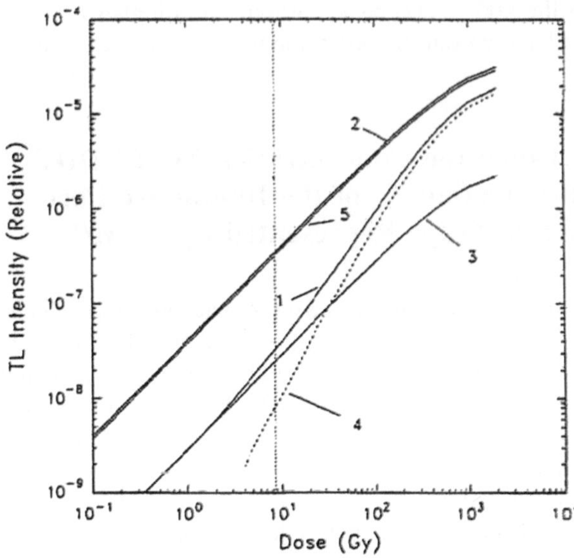

Fig. 6.11 Separation of linear and superlinear parts of the TL intensity growth curve of peak 5 of LiF:Mg, Ti (TLD-100); *curves 1* and *2* are the same as *curves 3* and *4*, respectively, of Fig. 6.10. *curve 3* is the extrapolation of the linear part of *curve 1* using (6.4) with $\alpha = 1.16 \times 10^{-3}$ Gy^{-1}. It gives intensity I_n produced by the noninteractive part n_n of the active trap population $(n_n + n_i)$. *Curve 4* is obtained by subtracting *curve 3* from *curve 1* and it represents the growth of TL intensity I_i due to the interactive part n_i of the active trap population. *Vertical dotted line* is drawn at $D = 8.6$ Gy [31, Fig. 6].

are 13 times of those in curve 3 at any dose, whereas those in curve 2 are 14 times. The value of $R(= I_i/I_s)$ at any given dose D may be found by using the values of I_i and I_s at the given dose. We have chosen to use the value of R at a low dose so that it may be applied to (6.27). It may be seen that at $\alpha D = 10^{-2}$ (which means at $D = 8.6$ Gy using $\alpha = 1.16 \times 10^{-3}$ per Gy), the value of R from Fig. 6.11, turns out to be about 2×10^{-2}. Substituting $R = 2 \times 10^{-2}$, $\alpha D = 10^{-2}$ and $\varepsilon/\alpha = 10^{-1}$ in (6.27), we arrive at the following relation

$$a = 0.49/(A_m/A_h - 0.049)$$

At $A_m/A_h = 0.1$, which is commensurate with ε/α value for LiF:Mg, Ti (TLD-100), value of a turns out to be 10. This value may be acceptable, at least within an order of magnitude. As discussed earlier the concentration of TDDT is expected to be greater than that of AT. Further reduction in A_m/A_h leads to a rapid increase in a till $A_m/A_h = 0.049$, beyond which a turns into negative. Sunta et al. [31] have attempted to verify the appropriateness of the model and the procedure adopted in determining the values of the parameters α, ε K and a by back-fitting the above obtained values of these parameters into the TL intensity growth (6.16). They find a good agreement between the computed growth curve and the actual growth curve of LiF:Mg, Ti TL phosphor.

References

1. S. Watanabe, *Research Report to FAPESP, Brazil, University of Sao Paulo.* (Institute of Physics, Sao Paulo, 1998)
2. N.A. Larsen, L. Botter-Jensen, S.W.S. McKeever, Radiat. Prot. Dosim. **84**, 87 (1999)
3. J.K. Srivastava, in *Proceedings of First International Conference on Luminescence and its Applications*, Luminescence Society of India Karaikudi TN India, 1992, Published as Luminescence: Phenomena, Materials and Devices ed. by R.P. Rao (Nova Science Publishers Inc., New York, 1992), p. 71
4. C.M. Sunta, Radiat. Effects **79**, 149 (1983)
5. A.R. Lakshmanan, K.G. Vohra, Nucl. Instrum. Methods **159**, 585 (1979)
6. A. Halperin, R. Chen, Phys. Rev. **148**, 839 (1966)
7. E.F. Mische, S.W.S. McKeever, Radiat. Prot. Dosim. **29**, 159 (1989)
8. C.M. Sunta, V.N. Bapat, S.P. Kathuria, in *Proceedings of the Third International Conference Luminescence Dosimetry (now called SSD)*, 11–14 October 1971. (Danish AEC Research Establishment Riso, Denmark, 1971), p. 146
9. M. Moscovitch, Y.S. Horowitz, J. Phys. Appl. Phys. **21**, 804 (1988)
10. R. Chen, X.H. Yang, S.W.S. Mckeever, J. Phys. D Appl. Phys. **21**, 1452 (1988)
11. A.R. Lakshmanan, R.C. Bhatt, S.J. Supe, J. Phys. D Appl. Phys. **34**, 1683 (1981)
12. B. Chandra, R.C. Bhatt, S.J. Supe, Nucl. Instrum. Methods **184**, 549 (1981)
13. J.K. Srivastava, S. Supe, J. RadiatProt Dosimetry **6**(1-4), 45–48 (1983)
14. M.J. Aitken, *Physics Reports, Archaeological involvements of Physics*, vol 40C, no 5. (North—Holland Publishing Company, Amsterdam, 1978), p. 291 (Fig. 2.4)
15. R. Chen, S.W.S. McKeever, Radiat. Meas. **23**(4), 667 (1994)
16. N. Suntharalingam, J.R. Cameron, Phys. Med. Biol. **14**, 397 (1969)
17. E.W. Claffy, C.C Klick, F.H. Attix, in *Proceedings of the Second International Conference Luminescence Dosimetry*, Gatlinberg, Tenn.(USAEC CONF-680920) (1968), p. 302
18. Y.S. Horowitz, M. Rosenkrantz, Radiat. Prot. Dosim. **31**, 71 (1990)
19. M. Rosenkrantz, Y.S. Horowitz, Radiat. Prot. Dosim. **47**, 27 (1993)
20. C.M. Sunta, Phys. Stat. Sol. **53**, 127 (1979)
21. C.M. Sunta, Phys. Stat. Sol. **37**, K81 (1970)
22. R. Chen, G. Fogel, Radiat. Prot. Dosim. **47**, 23 (1993)
23. E.T. Rodine, P.L. Land, Phys. Rev. **134**, 2701 (1971)
24. N. Kristianpoller, R. Chen, M. Israeli, J. Phys. D Appl. Phys. **7**, 1063 (1974)
25. C.M. Sunta, E.M. Yoshimura, E. Okuno, J. Phys. D Appl. Phys. **27**, 852–860 (1994)
26. C.M. Sunta, E.M. Yoshimura, E. Okuno, J. Phys. D Appl. Phys. **27**, 1337–1340 (1994)
27. R. Chen, G. Fogel, C.K. Lee, Radiat. Prot. Dosim. **65**, 63–68 (1996)
28. R. Chen, S.W.S. McKeever, *Theory of Thermoluminescence and Related Phenomena* (World Scientific, Singapore, 1997), p. 180
29. C.K. Lee, R. Chen, J. Phys. D Appl. Phys. **28**, 408 (1995)
30. C.M. Sunta, E.M. Yoshimura, E. Okuno, Radiat. Meas. **23**, 655–666 (1994)
31. C.M. Sunta, E. Okuno, J.F. Lima, E.M. Yoshimura, J. Phys. D Appl. Phys. **27**, 2636–2643 (1994)
32. A.R. Lakshmanan, R.C. Bhatt, S.J. Supe, J. Phys. D Appl. Phys. **14**, 1683 (1981)
33. B. Chandra, R.C. Bhatt, S.J. Supe, Nucl. Instrum. Meth. **184**, 549 (1981)
34. A.R. Lakshmanan, R.C. Bhatt, K.G. Vohra, Phys. Stat. Sol. **53**, 617 (1979)

35. Niewiadomaski T., Comfrontation of Thermoluminescence models in Lithium fluoride with experimental data, Institute of Physics Krakow (Poland) Report Dec (1976)
36. C.M. Sunta, E.M. Yoshimura, E. Okuno, Phys. Stat. Sol. **142**, 253 (1994)
37. A.R. Lakshamnan, B. Chandra, R.C. Bhatt, J. Phys. D Appl. Phys. **15**, 1501 (1982)
38. M.R. Mayhugh, R.W. Christy, M.N. Johnson, J. Appl. Phys. **41**, 2968 (1970)
39. G.C. Critenden, P.D. Townsend, S.E. Townsend, J. Phys. D Appl. Phys. **7**, 2397 (1974)
40. L.V.E. Caldas, M.R. Mayhugh, T.G. Stoebe, J. Appl. Phys. **54**, 3431 (1983)
41. J.L. Landreth, S.W.S. McKeever, J. Phys. D Appl. Phys. **18**, 1990 (1985)
42. B. Chandra, A.R. Lakshmanan, R.C. Bhatt, J. Phys. D Appl. Phys. **15**, 1803 (1985)

Chapter 7
Photo-Transferred Thermoluminescence

Abstract This chapter deals with transfer of TL intensity from one glow peak to another under exposure to light. The transferred intensity is called photo-transferred TL (PTTL). This phenomenon is observed almost in all thermoluminescent materials, usually showing a transfer from higher temperature glow peak to those of lower temperatures. In this Chapter transfer is shown from a glow peak to lower as well as to higher temperature glow peaks. This chapter also describes the effect of sample temperature on the transfer efficiency for incident light. This phenomenon is used to find the energy levels of the donor trap. Using the monochromatic incident light for photo-transfer, optical trap depth is found for the donor trap. Studies describing the dependence of transfer efficiency on sample temperature have possibly not been given in earlier books of this field. These studies are used to propose a model for photo-transfer in mineral CaF_2. Application of PTTL in cumulative dose measurement and retrospective dosimetry is described briefly. Finally the topic of photo-bleaching is taken up. It is an important subject in sediment dating. It is shown that in a multi-peak glow curve photo-bleaching and photo-transfer run together. On prolonged light exposure, all glow peaks, donors as well as accepters attain a low and nearly stable level. At this stage most of the traps are empty and correspondingly the population of the recombination centres also is negligibly small. As a result the available option for the photo-excited electrons from any of the traps is simply to get retrapped into any of the vacant traps. In other words a to and fro transfer between the traps continues almost indefinitely. This leads to a nearly unbleachable TL signal which is used as the zero level in the TL dating of geological sediments.

7.1 Definition and Methodology

Transfer of light intensity from one thermoluminescence (TL) glow peak to another under exposure to the light is termed as photo-transfer of thermoluminescence (PTTL). The phenomenon is distinctly different from the primary

C. M. Sunta, *Unraveling Thermoluminescence*,
Springer Series in Materials Science 202,
DOI: 10.1007/978-81-322-1940-8_7,

induction of TL by ionizing radiations. In the case of primary induction of TL, electrons are made free from the valence band. Liberation of electrons from the valence band requires more than the band gap energy of the phosphor. The band gap energies of various TL phosphors lie in the range of about 7–12 eV. It means that direct induction of TL by UV light would require UV rays of vacuum UV region. In terms of wavelength λ the upper limit that can release an electron from the valence band is given by

$$\lambda = 12{,}400/E \tag{7.1}$$

where λ is in Å and E is the band gap in eV. In PTTL the λ values of light used are generally much greater ($\sim$2,500–5,000 Å).

The phenomenon of PTTL is a three-step process. In the first step electrons and holes are made free from the valence band to induce TL glow peaks as described earlier in Chap. 3. In the second step some of these glow peaks are erased by heating the sample to a pre-decided temperature. In the third step light is shined on the sample to once again produce the glow peaks erased in the second step.

An illustrative example of the PTTL phenomenon may be seen in the blue-green mineral fluorite [1]. A typical glow curve of this mineral above the room temperature is shown in Fig. 7.1a. As many as five prominent glow peaks may be seen from room temperature to 600 °C. When a sample of this mineral is given a high dose ($\sim 10^3$ Gy) and heated up to 400 °C, the glow peak appearing at about 500 °C remains residual in the sample. It is shown by the dashed line in Fig. 7.1b. Subsequent exposure of such a sample to UV light causes the re-induction of the lower temperature glow peaks. These are shown by the solid line. These glow peaks in the high dose irradiated sample had been wiped out during heating up to 400 °C. Such re-induced glow peaks may in fact be seen all the way below the room temperature up to -150 °C [2]. In Fig. 7.1 it may be seen that the UV exposure also causes some reduction in the residual TL (RTL) glow peak of 500 °C. When the sample is heated up to about 700 °C after a high dose, no TL glow peaks are induced by UV light exposure, nor could any RTL be seen in the heated sample. The obvious conclusion from these observations is that there is a transfer of TL from the high temperature residual peaks to the lower temperature peaks. This parent-offspring relation between the RTL peak and the re-induced peaks by light exposure is further supported by the linear relation between the intensity of the RTL peak and that of the re-induced peak shown in Fig. 7.2. This figure shows that the intensity of the re-induced peak appearing at about 265 °C is directly proportional to the intensity of the RTL peak. In this experiment varying intensities of RTL peak are produced by giving different doses (1–10^3 Gy) to a 650 °C heated sample and then heating it up to 400 °C. UV exposures of 1 min duration were given to equal aliquots of the samples which have been given different doses. The UV source is a 125 W Philips Black-light lamp placed at a distance of 25 cm from the sample. Apart from showing the donor–acceptor relation between the RTL peak and the PTTL peaks, this experiment tells that the charge carriers trapped at the RTL peak and in the peaks induced by light exposure are of the same kind of charge. Glow peaks which

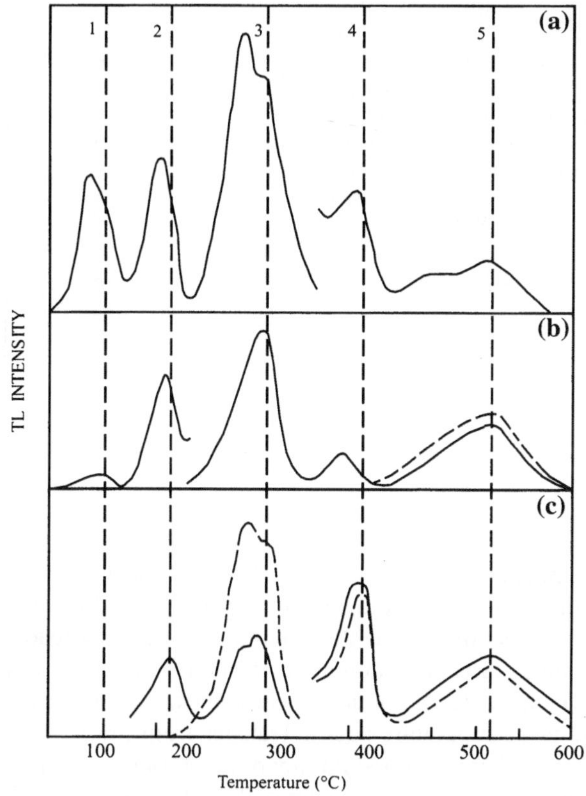

Fig. 7.1 Photo-transfer of glow peaks, **a** TL glow curve of a mineral CaF_2 sample from RT to 600 °C after a gamma dose of 50 Gy. Prominent glow peaks are marked 1–5. **b** *Dashed line* is the residual glow curve of a 10^3 Gy gamma irradiated sample subsequently heated at 400 °C. The *continuous line curve* is the PTTL glow curve after UV exposure of this sample. **c** *Dashed line curve* is for the sample of (**a**) after heating at 150 °C to erase glow peaks of temperature below 200 °C. On exposure of this sample to yellow light ($\lambda \geq 500$ nm), the TL induced is shown by the *continuous line*. (Derived and modified from [1], © with permission from John Wiley and Sons)

are not produced in the PTTL experiment but are seen only in the β and γ exposed samples may involve defect centers in which the nature of charge carrier may be different from those of the RTL and the observed PTTL peaks. For example in this sample of CaF_2 the major component of peak 3 (third from room temperature) appears missing in the PTTL glow curve. Only the weaker component appearing at a slightly higher temperature is seen as the prominent peak of the PTTL glow curve. In the blue-green CaF_2 mineral sample the transfer of TL may be seen also from the lower temperature to higher temperature peaks [1]. This is seen by exposing a sample to a non-saturating gamma dose of about 10 Gy so that the glow peaks are not saturated by ionizing radiation exposure and there is scope for their increase. This irradiated sample is then heated up to 200 °C. This leaves the glow peaks appearing above this temperature as residual in the sample. Subsequent exposure to yellow light induces the lower temperature glow peaks which had been wiped out during the heating and also enhances the intensity of the residual peaks of temperature 350 °C and above. The intense glow peak appearing at about 220 °C is reduced in intensity (see Fig. 7.1c). The conclusion may simply be drawn that TL intensity has been transferred from the middle peak of 220 °C to both lower and higher temperature peaks. While the transfer of TL from higher to lower

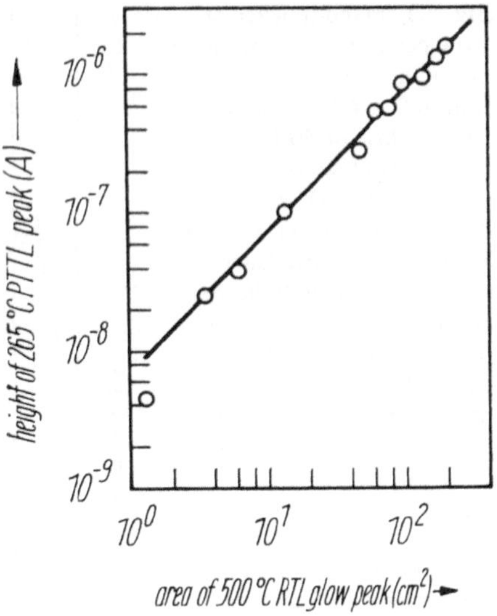

Fig. 7.2 Intensity of PTTL glow peak of 265 °C versus intensity of RTL peak of 500 °C. UV exposures are of 1 min duration using 125 W Philips Black light lamp at a distance of 25 cm (From [2]). Reproduced with permission from John Wiley and Sons Copyright © 1979 Wiley-VCH Verlag GmbH & Co. KGaA

temperature peak is easy to observe and can be seen in almost any of the TL samples, transfer from lower to higher temperatures is difficult to observe, because it involves the detection of a small added intensity to an already existing glow peak of higher temperatures. It also requires a careful selection of the incident light wavelength, which would bleach the concerned lower temperature peak, but does not bleach the higher temperature peaks. Transfer from lower to higher temperature glow peaks has not been reported for any other phosphor except for mineral CaF_2 shown in the above figure. Photo-transfer of TL intensity from higher to lower temperature glow peaks is usually affected by exposure of the sample to a UV source such as quartz envelope mercury discharge lamp. Such mercury lamps have strong UV emission lines at 365 and 254 nm wavelengths. In fact any light which causes reduction in the intensity of a glow peak (photo-bleaching) can in principle produce photo-transfer from that peak. Usually the higher energy components of white light (blue, violet, and UV) cause bleaching of most TL glow peaks. It is because of this that the laboratories where samples are handled for TL or OSL work use yellow or red light for general illumination to avoid photo-bleaching of TL. For transfer from lower to higher temperature peaks it is expected that the incident light photon energy should be lesser than that for transfer from higher to lower temperature peaks. This is because, generally we expect the trap energies to increase with the temperature of the glow peak (refer Sects. 4.2 and 4.3.1 in this book).

7.2 Phenomenological Understanding

PTTL glow curves of a number of $CaSO_4$, CaF_2 and $CaCO_3$ phosphors for transfer of TL from higher to lower temperature glow peaks, have been given by Sunta [3] in a review article on calcium based TL phosphors. A phenomenological understanding of PTTL may be given as following: When some glow peaks are left residual in a sample after partial readout of the glow curve, it means that the traps corresponding to the residual peaks continue to remain filled. These trapped charges may be ejected out by the incident light photons provided the photon energy is greater than the binding energy (trap depth) of the charge to the given trap. The photon energy which is just sufficient to eject the charge is called the optical activation energy of the trap. The process is thus like photo-electric effect. The ejected out charges are free to move in the delocalized band (electrons in conduction band and holes in valence band) and these fall back to wherever they find a vacant place. When they fall into the recombination centers, luminescence may be observed. It is called *optically stimulated luminescence* (*OSL*). Alternatively these free carriers may get retrapped in any of the vacant trapping sites. The traps of the lower temperature peaks, being all vacant because of prior heating of the irradiated sample during partial readout of the glow curve, the free charges may mostly get retrapped into these locations. Additionally some may fall into the traps of higher temperature peaks if some of these are vacant. This will particularly happen if the radiation dose given for TL induction is smaller than the saturation dose. Thus the population of the filled traps of higher temperature may also be increased as seen by the increase in the intensity of the higher temperature glow peaks shown Fig. 7.1c. Some of the charge carriers may also be recaptured in the vacant traps of the same type from where these were ejected out. Thus it is seen that the PTTL process results in redistribution of the optically released carriers. The process is illustrated by the scheme shown in energy level diagram of Fig. 7.3.

7.3 Use of PTTL in Basic Studies of TL Materials

7.3.1 Energy Levels of the RTL Traps of the Mineral CaF_2

(a) Optical trap depth—To find the optical trap depth of the donor traps we may study the dependence of the PTTL intensity on the photon energy of the light used for inducing the photo-transfer. Such studies require elaborate experimental arrangement. The essential requirement is that the photon flux of light at different wavelengths should be known in order to expose the samples to known incident photon fluxes for inducing PTTL. For this purpose, we should have a standard lamp with known spectral distribution and a monochromator with known spectral efficiency at different wavelengths. Alternatively if a standard lamp is not available one may standardize a light source such as a deuterium lamp using a monochromator and a PM tube whose spectral response is precisely known. The relative photon flux at the exit slit for each wavelength setting is determined before light exposures are carried out to induce the PTTL in the given sample. The results shown in Fig. 7.4 are for the PTTL glow peak of 265 °C induced in a mineral CaF_2 sample heated at 400 °C after a high dose exposure. These show distinct peaks at light wavelengths ~366 nm and 231 nm which in terms of photon energy mean 3.38 and 5.36 eV

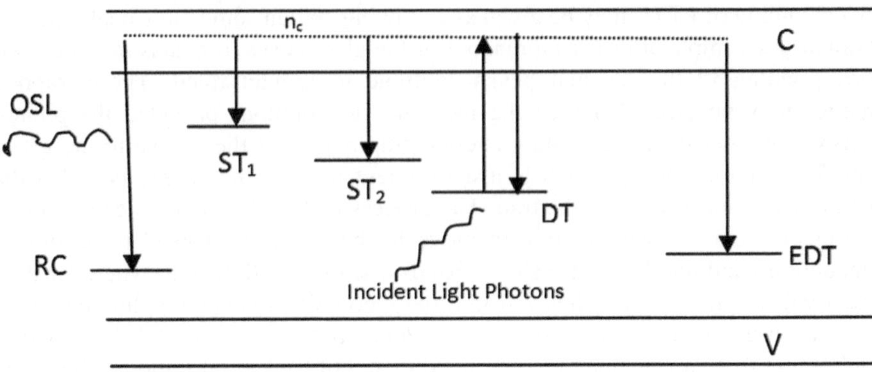

Fig. 7.3 Phenomenological model of phototransfer of TL. DT is the donor trap, ST_1 and ST_2 are shallower acceptor traps which are empty when light exposure is started. EDT is the extra deep trap at a further deeper level than the residual donor trap DT. The incident light photon energy is such that it bleaches the DT traps but does not affect EDT. RC is the recombination centre, transition into which may produce OSL

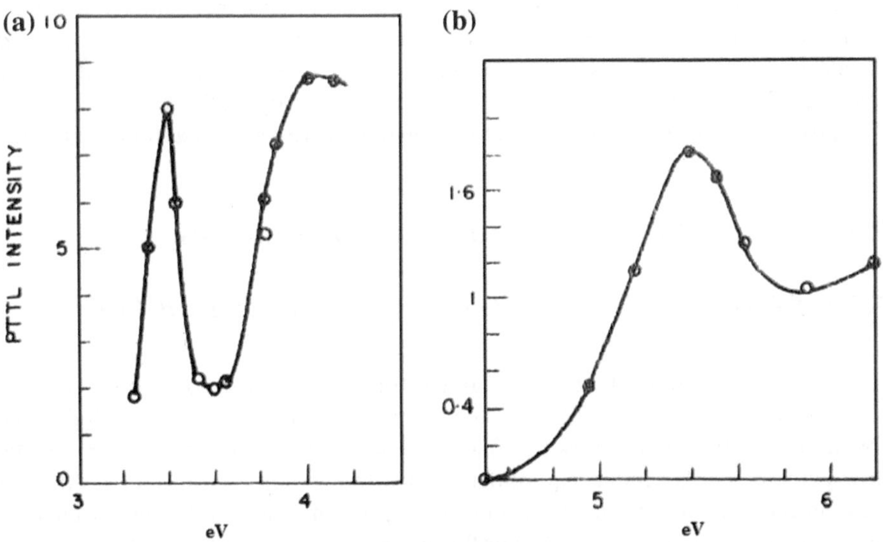

Fig. 7.4 PTTL intensity (relative) per light photon fluence incident on the sample. Curve (**a**) is for PTTL intensity produced by transfer of TL from RTL peak of 500 °C and curve (**b**) for PTTL intensity produced by transfer from the RTL peak of 650 °C, [3, Fig. 13] © Reproduced with permission of Oxford Univ. Press

respectively. These energies represent the optical activation energies of the traps of RTL peaks of 500 and 650 °C respectively. Curves (a) and (b) are for samples heated at 400 and 550 °C respectively after a dose of 10^4 Gy. The 650 °C glow peak is very faint [2] and is measurable only after high β/γ dose and requires a carefully selected filter combination which includes heat and infra red filters placed over the sample heater [4]. It is noteworthy that the above activation energy values are much higher than the thermal activation energies of these peaks measured by initial rise method. The thermal activation energy values of these traps have been found as 2.04 and 2.99 eV respectively [5, 6]. Difference between the thermal and optical activation energies is attributed to the mutual displacement of the configuration coordinate curves of the traps in the ground and the upper states [7].

(b) Excited state energy levels of the RTL traps—The intensity of PTTL has been found to increase with the temperature of the sample during light exposure to induce the PTTL. This effect and its interpretation was first reported by Sunta [5]. The temperature dependence of PTTL induction may be studied by exposing the sample to a given light source at different temperatures [3, 5]. The results presented here are for the same mineral CaF_2 sample as used in para (a) above. Short duration light exposures (depending on the intensity of the light source used) of fixed duration are given at different sample temperatures such that there is no detectable bleaching of RTL peak of 500 °C, which is the donor trap in this case. Same sample may then be used repeatedly for exposures at different temperatures. The intensity of the 265 °C PTTL peak, which is most prominent among the PTTL peaks (see Fig. 7.1b) is plotted in Fig. 7.5 for exposures with 365 and 254 nm light. Normalized intensity values of the same glow peak induced by beta or gamma ray exposures of same duration and at same sample temperatures also are plotted for comparison. It is observed that the intensity of the PTTL peak increases with the temperature of the sample until the temperature approaches that of the peak maximum (T_m). At that stage the thermal bleaching overtakes the enhancing effect of temperature. On the other hand there is no increase in the TL intensity in the case of the samples exposed to beta or gamma rays at different temperatures. They only show the thermal bleaching effect. The extent of thermal bleaching of the PTTL peaks at different temperatures may therefore be found by using the TL of the beta or gamma irradiated samples. The enhancing effect of sample temperature on PTTL induction appears interesting because the role of temperature in general is to bleach out the TL. The temperature dependence of PTTL induction was studied by Sunta [5] for a CaF_2 sample from Ambadungar Gujarat, India. Subsequently, this work was extended by Bailiff et al. [8] to MBLE – Super S sample of CaF_2 from Belgium. The interpretation of the phenomenon has been given as follows [5]: The RTL trap may have intermediate energy levels between its ground state and the free-state. These intermediate energy states may be either closely above the ground state of the trap or closely below the free-state. The thermal energy helps in crossing these small energy gaps before the incident light photon is able to make the trapped charge free from the RTL trap. Assuming that this small energy gap is ΔE, the probability for attaining energy ΔE at any temperature T may be given by Boltzmann's relation for population in thermally excited states. The probability for transfer of a charge carrier from a RTL trap to other empty traps via the delocalized band and consequently the observed PTTL intensity I, would therefore increase with temperature according to the Boltzmann's relation viz

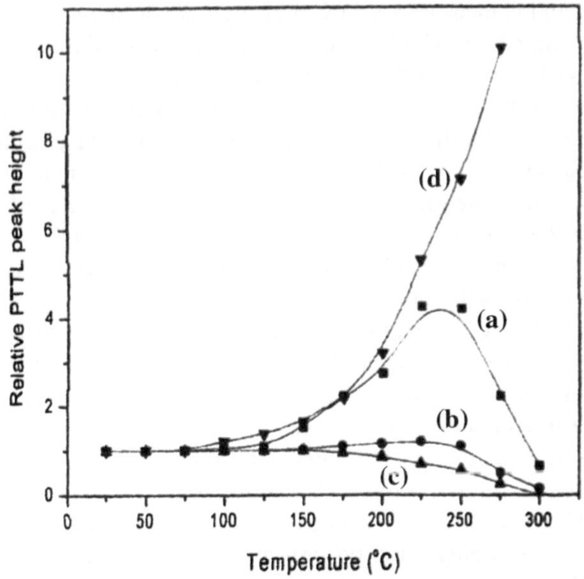

Fig. 7.5 Effect of heating on photo-transferred intensity of the prominent PTTL peak of 265 °C of mineral fluorite (*a*) by 366 nm UV light exposure, (*b*) by 254 nm UV light exposure, (*c*) intensity of same TL peak by Ra gamma exposure. All intensity values are plotted as a function of sample temperature during exposures. All exposures are of 15 s duration. (*d*) Values of the observed intensity (*I*) of curve (*a*) corrected for thermal drainage during exposure by 366 nm UV light. The drainage factor is calculated from curve (*c*), since gamma induced TL is not enhanced by temperature

$$I \alpha \exp\left(-\frac{\Delta E}{kT}\right) \tag{7.2}$$

where k is Boltzmann's constant and T is the sample temperature. The observed PTTL intensities shown as curve (a) of Fig. 7.5 were corrected for the thermal bleaching during light exposure at elevated temperatures. The correction factors for this were calculated from curve (c) of Fig. 7.5, which is for β/γ irradiated samples. The corrected intensities of curve (a) are shown as curve (d) of this figure. When these intensities I are plotted as $ln\, I$ versus $1/T$, a straight line is obtained the slope of the line being $\Delta E/k$. The value of ΔE can be calculated from this. The graph using the data of curve (d) of Fig. 7.5 is found to consist of two straight lines shown in Fig. 7.6. The ΔE values correspondingly are found to be 0.32 and 0.08 eV. These are marked as E_1 and E_2 in Fig. 7.6. A tentative energy level diagram for the glow peak of 500 °C which is the RTL donor peak in this case is shown in Fig. 7.7. It is to be noted that the PTTL induced by 254 nm light is not so much enhanced by the rise in sample temperature (see curve (b) in Fig. 7.5). It may be because the light of this wavelength may cause direct transition into the conduction band. This inference is supported by the data plotted in curve (a) of Fig. 7.4 which has a plateau at photon energy above about 4 eV. In this case the plateau may appear because the photons

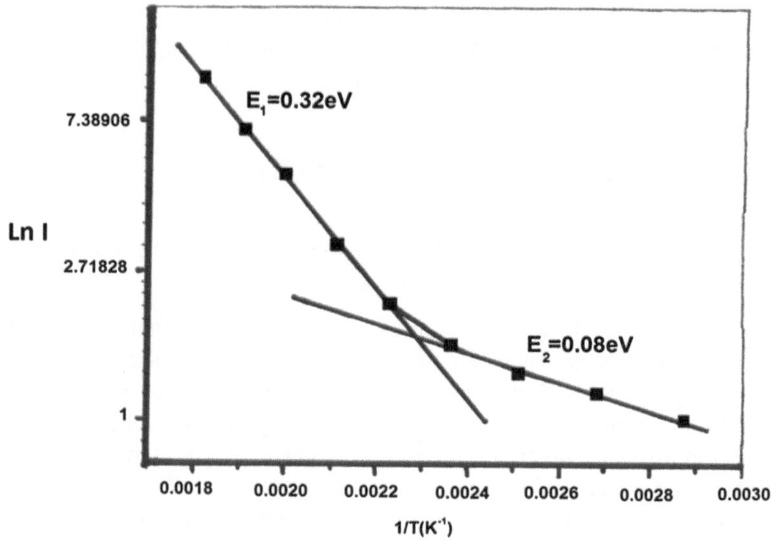

Fig. 7.6 *ln I* versus *1/T*. The *I* values are taken from curve *d* of Fig. 7.5. The y axis markings are in cycles of *e*. ($e = 2.71828$)

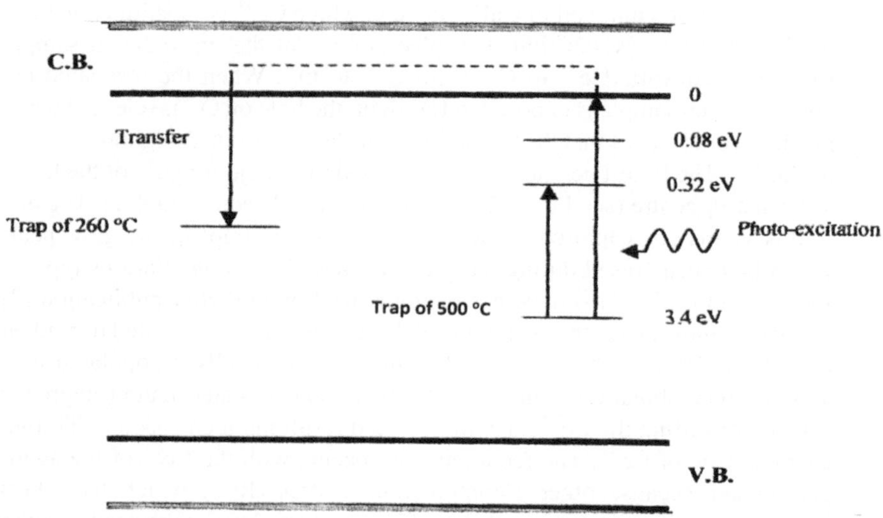

Fig. 7.7 Proposed energy level diagram for temperature dependence of photo-transfer of TL from the residual filled trap of 500 °C glow peak to the glow peak of 260 °C

above this energy may ionize the RTL trap of 500 °C by direct transition into the delocalized band. The PTTL in such a case would not be enhanced by the temperature during the light exposure. The above method of determining ΔE values is straight forward if the RTL consists of a single energy trap. If the RTL trap has multiple components, one may not get a straight line for the graph of *ln I* versus. *1/T*, since more than one ΔE values may be present.

7.3.2 Mechanism of PTTL in Mineral CaF_2

A model for the mechanism of PTTL in this material has been proposed by Sunta [2]. It is essentially based on the following facts:

(1) The intensity of PTTL peaks is directly proportional to that of the RTL peak
(2) The emission spectra (wavelengths) of all the PTTL peaks are same as that of the RTL peak (500 °C glow peak). These spectra consist of a broad band with maximum at about 380 nm. These wavelengths have been identified as emission from a substitutional Ce^{3+} (in place of Ca^{2+}) perturbed by O^{2-} (substituted in place of F^-) in CaF_2 lattice [9]. When a β/γ irradiated sample is heated up to 400 °C to erase all the glow peaks appearing below that temperature, the traps corresponding to the glow peaks appearing below 400 °C are all emptied in the process of heating and the peak of 500 °C remains as the main RTL peak. The model proposes that this RTL peak is due to an aggregate electron-hole centre. This aggregate centre is illustrated in Fig. 7.8 at position 1 in the form of a unit cell of CaF_2 lattice in which Ce^{3+} is substituted in place of Ca^{2+} and O^{2-} is substituted in place of F^- in the unirradiated sample. Irradiation converts Ce^{3+} to Ce^{2+} and O^{2-} to O^-. When the irradiated and 400 °C heated sample is exposed to UV light, the hole on O^- is released which in other words means that the O^- picks up an electron from a neighboring F^- of the lattice. The F^0 left behind may get attached to an adjoining F^- of the lattice to form a V_k centre (see Fig. 2.2 in Chap. 2) which becomes mobile. The hole of this V_k may be captured in the lower temperature traps of the glow peaks which had been erased during the previous 400 °C heating. Various types of such hole traps in CaF_2 have been summarized in an earlier publication [3]. The PTTL glow peaks thus get induced. When this sample is heated to read out the PTTL glow, the holes released from the photonically repopulated traps undergo recombination with the Ce^{2+} of the above stated parent aggregate center, converting the Ce^{2+} back to Ce^{3+} and resulting in emission of TL light characteristic of Ce^{3+}. The recombination occurs with the Ce^{2+} of the aggregate center because other recombination centres (RCs) which have been identified as different rare earth ions [3] may have been exhausted together with the lower temperature glow peaks of $T_m < 400$ °C during the previous 400 °C heating. When the sample is heated further to read out the residual 500 °C glow peak (see the arrow from the position 2 to 1 in the figure), the available recombination centre again is the same Ce^{2+} center. Thus the emission spectrum of the 500 °C RTL glow peak is same as that of the PTTL peaks. This stepwise sequence in this PTTL experiment is represented schematically in Fig. 7.8.

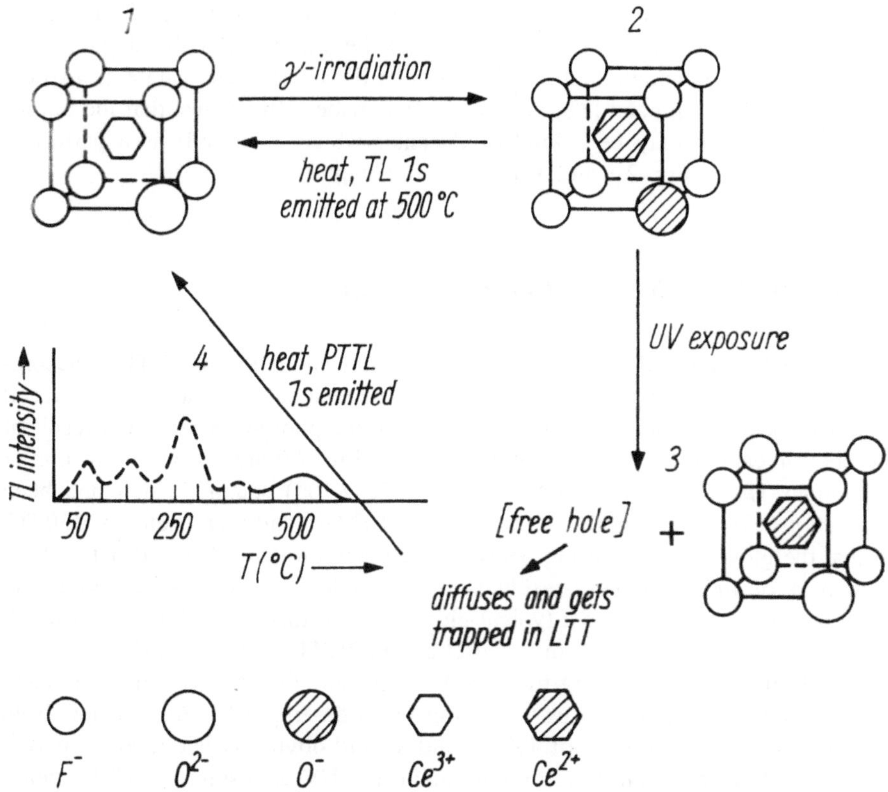

Fig. 7.8 A model for PTTL for mineral CaF_2. Stage 1: Centre responsible or RTL before irradiation. Stage 2: The same after irradiation. It is an electron-hole aggregate centre. Stage 3: UV irradiation of stage 2 sample after heating at 400 °C liberates a hole (O^{-}–UV → O^{2-} + hole) and centre of stage 2 sample is now left only with the electron centre (Ce^{2+}). The free hole diffuses and gets retrapped into the vacant traps of lower temperature glow peaks. Stage 4: The sample is now heated to read the PTTL, the glow peaks of PTTL shown by dashed part of the glow curve are produced. The solid part is the residual TL left after 400 °C heating of stage 2 sample, which too is emitted now. With heating the defect centre returns to its original state. (From [3]), Reproduced with permission from John Wiley and Sons, Copyright © 1979 Wiley-VCH Verlag GmbH & Co. KGaA

7.3.3 Study of PTTL in LiF:Mg, Ti (TLD-100)

Similar studies as discussed above in respect of mineral CaF_2 have been carried out for the well known phosphor LiF:Mg, Ti (TLD-100) by Sunta and Watanabe [10]. They found that the excitation spectrum of the PTTL peak, for which the residual donor peak is at 400 °C, has maximum value at 5.25 eV. This means that the optical activation energy of the 400 °C glow peak is 5.25 eV. In this LiF sample also the PTTL intensity *I* increases with sample temperature during UV exposure. The plot of *ln I* versus *1/T* shows a single straight line. The slope of which yields a ΔE value of 0.12 eV. This means that the trap of 400 °C glow peak in this material has an excited state at 0.12 eV below the delocalized state (conduction band). It is known that the F center of LiF has an excited state at 0.16 eV under the conduction band [11]. Sunta and Watanabe [10] suggest that the trap of 400 °C glow peak may be a kind of *Z* center which essentially is a perturbed *F* type center (see Fig. 2.1 and Table 2.1a in Chap. 2).

7.4 Photo-Transfer and Photo-Bleaching

As seen in Fig. 7.1 and explained in Sect. 7.2, the induction of PTTL is accompanied with a reduction in the TL intensity of the residual TL peak, which we call the donor peak. In other words we may say that the photo-transfer is the result of the photo-bleaching of the residual TL peak. In Fig. 7.9 are shown the results of continuous UV exposure of a LiF:Mg, Ti (TLD-100) sample which had been exposed previously to a high gamma dose ($\sim 10^3$ Gy) and then heated at 350 °C, so that the glow peak appearing below that temperature are bleached out and the glow peak of 400 °C remains residual in the sample. This peak is marked as No. 10 in the inset, which shows a typical glow curve of this TLD phosphor. The TL peaks marked as 2, 3, 5 and 7 in the inset are the PTTL peaks. It may be seen that the PTTL glow peaks resulting from the UV exposure first rise to a maximum and then start falling. The rise is due to the transfer of the trapped charge carriers from the traps of the residual peak of 400 °C and the fall obviously is due to the photo-bleaching. After a certain time on continuing the UV exposure, the PTTL peaks are seen to attain a steady level of intensity. The residual peak of 400 °C marked 10 which has an initial fall as expected, also appears to attain a steady level. It has been explained by Sunta et al. [12] that this behavior of the residual and the PTTL peaks occurs because of the to and fro transfer of the trapped charge carriers between the traps of these glow peaks under the UV light exposure. In other words it is the effect of untrapping and retrapping of the charge carriers under the continuous light exposure. Apparently there is no significant loss of charge carriers by recombination transitions. Thus further bleaching of the TL under U V exposure becomes insignificant. The near absence of recombination transitions may occur due to the fact that when most of the traps of the PTTL as well as RTL are rendered vacant by the light exposure, the concentration of RCs also becomes insignificantly low (because of the charge balance between the total number of filled traps and available number of the RCs). Due to the abundance of various types of vacant trapping sites and the paucity of recombination centres, nearly all of the untrapped carriers during continued light exposure take the retrapping route for relaxation. Thus further exposure to light is unable to reduce the trapped charge population to zero in a finite time. The bleaching behavior like the one seen in Fig. 7.9 has also been recorded for mineral CaF_2 [1] and many terrestrial and ocean sediments [13, 14, 15] which are naturally exposed to light and other environmental stresses during transport before deposition as sediment. In sediment

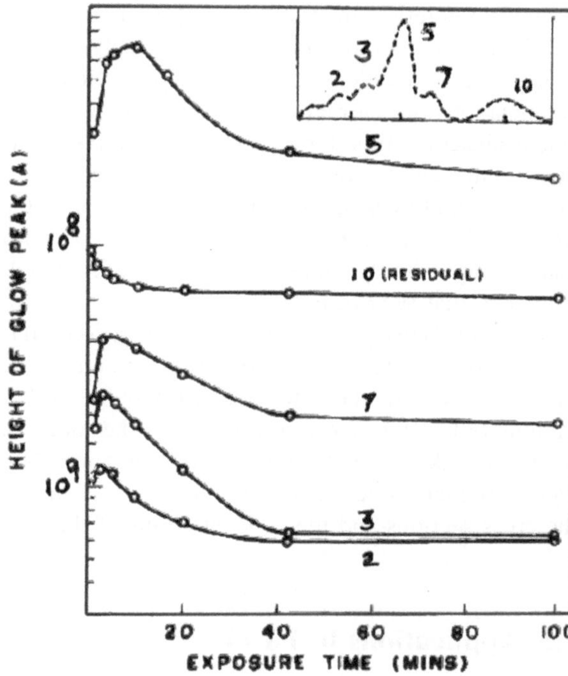

Fig. 7.9 Photo-transfer and photo-bleaching of glow peaks of LiF:Mg, Ti (TLD-100). Curve No. 2, 3, 5, 7 and 10 represent the glow peaks of same number in the glow curve shown as inset

TL dating the unbleachable part is an important parameter, since it provides the 'zero level' of TL for transported and buried sediment dating.

The above discussed behavior of photo-bleaching and photo-transfer has been given a theoretical footing by Levy [16]. He described this behavior as the second order kinetics phenomenon in the TL of the minerals because of the involvement of retrapping. This pattern of photo-transfer and photo-bleaching appears to be universal in all TL emitting materials including in those whose kinetic order is well known to be of first order. The case of LiF:Mg, Ti described above is an outstanding example. Not-withstanding the extensive retrapping during the photo-bleaching, the glow peaks of TL samples may not show the typical characteristics of second order or general order kinetics, namely the shift of the glow peaks to higher temperatures with bleaching time. The glow curves of LiF:Mg, Ti (TLD100) whose photo-bleaching is shown in Fig. 7.9, were specifically recorded after various bleaching times and were found to retain the same peak position ($T_m \sim 200$ °C) [17] showing that this material belongs to the FO kinetics. The question why FO kinetics prevails in the TL of real materials is already discussed in Sect. 3.7. Using the theoretical models it is shown in Sect. 3.5 that the glow peaks attain first order kinetics at low concentration of the traps in the physical models involving thermally disconnected deep traps. Thus designating the TL bleaching behavior to second order kinetics may not be relevant.

In his theoretical treatment Levy [16] assumes the traps of some glow peaks as pre-filled and some others as initially empty when the photo-bleaching is begun. Using the rate equations for the untrapping and the retrapping processes, numerical computations are carried out for determining the profiles of filled concentration versus bleaching time for each of the traps. As expected the filled concentration of the initially empty traps shows an increase in the beginning before reaching a maximum and then falling. The concentration of the initially filled traps shows a steady fall. After a while the decrease in concentration for both types of traps namely the initially empty and the initially filled ones, becomes insignificant similar to that seen in Fig. 7.9.

Using the rate equations in a similar way as used by Levy [16], one may attempt to carry out fitting of the photo-transfer cum photo-bleaching pattern shown in Fig. 7.9. The parameters to be fitted may be cross sections and the total concentrations (N) of the five types of traps. The total concentration of the donor trap (residual peak 10) and the incident light flux may be taken as known using realistic values for them (refer Sect. 5.5 in this book). This exercise may yield the values of the cross sections and the concentrations of the traps.

7.5 Applications of PTTL

7.5.1 Cumulative Dose Measurement of Radiation Workers

The heating temperature of TL dose meters during TL readout to assess the dose received by the radiation workers does not normally exceed 350 °C. This leaves the TL above this temperature as residual. The intensity of these residual peaks is weak. However, when these dose meters are reused, the TL signal of the residual peaks builds up with the addition of dose during each period of reuse. This accumulated signal is still weak for direct measurement. Also the heat signal from the TLD reader pan interferes with TL signal from the residual peak. The use of PTTL technique comes handy to measure residual peak intensity, since the PTTL intensity is stronger by orders of magnitude in comparison to the residual glow peak. Figure 7.2 shows that the PTTL intensity is directly proportional to the residual peak intensity. One may thus estimate the accumulated dose from the PTTL. In the case of mineral CaF_2 TLD, the residual peak of 500 °C has been found to grow linearly with dose [9]. This means that the PTTL intensity too would be directly proportional to the accumulated dose.

7.5.2 Retrospective Dosimetry

Retrospective dosimetry means to assess the radiation dose to which a population or an individual might have been exposed after an accidental exposure to radiation. The well known example of an accident involving radiation exposure to population is the Chernobyl reactor accident in former USSR in which huge amount of radioactivity got released into the atmosphere. It was carried by the wind over large parts of Europe resulting in radiation exposure of varying degrees to the population groups in European countries. Estimates of radiation dose received by them had to be obtained by the retrospective method of dosimetry.

The principle of retrospective dosimetry is essentially the same as discussed above in the case of the cumulative dose measurement for the occupational radiation workers. For retrospective dosimetry one may use TL sensitive materials from house-hold items such as ceramic sanitary ware, jewels in watches or on personal ornaments. One may directly measure the TL of such materials or if a long period has passed during which TL of the usual temperature range of measurement might have faded, one may use the PTTL technique since the high temperature residual TL peaks may remain unaffected over a long period after the radiation exposure. TL measurement of high temperature glow peaks directly is beset with problems of thermal quenching at high temperatures and the thermal glow from the heater device. One may therefore adopt the PTTL method to assess the dose.

7.5.3 Ultraviolet Measurement

Ultra-violet (UV) radiations are injurious to the eyes and the skin. The PTTL technique may be employed to measure the UV intensity. A TL phosphor exposed to nearly saturation dose and partially heated so as to leave the high temperature glow peaks as residual may be used to measure the UV intensity. This pre-dosed and partially heated TL phosphor may be used in the form of a thin pellet or a chip. If the phosphor is available only in the powder form, it should be finely deposited on a metal chip using an adhesive like silicone resin. Pellets or chips may be placed on the uncovered parts of the body like the hands or face of the individual working near the UV sources. The advantage of this method of UV exposure monitoring over the conventional light measuring devices is that these may be used to measure the integrated UV dose over a working period of the individual workers whereas the conventional devices measure the instantaneous intensity. Figure 7.10 shows the PTTL intensity as a function of integrated UV exposure for the UV dose meter made of mineral CaF_2 powder. It may be seen that the intensity of the PTTL glow peak ($T_m \sim 265°C$) is directly proportional to the integrated UV exposure. One has to ensure that the residual donor peak does not fade significantly by UV exposure nor does the induced TL peak used for measuring the dose gets significantly bleached during the period of measurement. A proper calibration is essential before such a device is put to use.

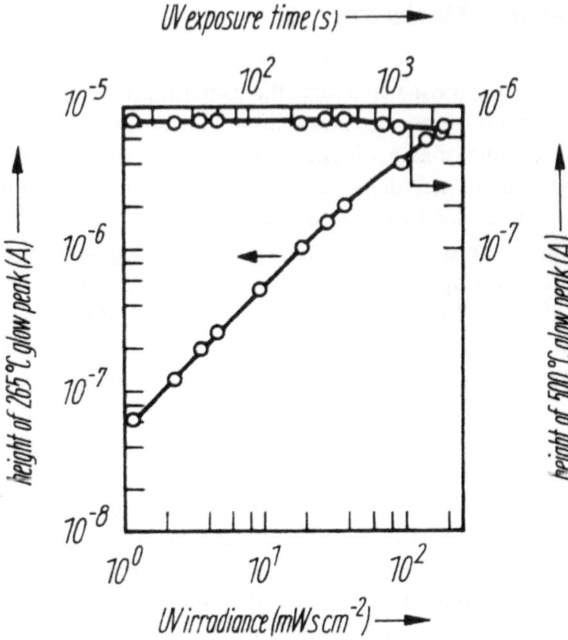

Fig. 7.10 PTTL intensity versus integrated exposure of the UV dosemeter made of mineral CaF_2 pre-exposed to about 10^3 Gy of gamma dose and then heated at 400 °C. The lower curve shows the intensity of the of the PTTL peak and the upper curve shows the intensity of the RTL peak of 500 °C. It may be seen that the RTL peak shows no significant fading until the UV exposure exceeds about 40 mWs cm^{-2}. The PTTL continues to grow linearly until the RTL peak does not show a significant fading. (From [3]), Reproduced with permission from John Wiley and Sons. Copyright © 1979 Wiley-VCH Verlag GmbH & Co. KGaA

7.5.4 Thermal Quenching Factor of High Temperature TL Glow Peaks

It is well known that the luminescence efficiency generally decreases with increase in sample temperature. This is called thermal quenching. The thermal quenching effect on individual glow peaks has been measured in two ways (1) by recording the radioluminescence or photoluminescence as a function of temperature [18, 19]. These methods may be used when the emission wavelengths are same in radio-luminescence and photoluminescence as in thermoluminescence. When thermal quenching is present, the radioluminescence and photoluminescence will be seen to decrease with the sample temperature. If thermal quenching is present in the temperature range of a composite TL peak, a distortion may appear in the shape of the peak (sharper fall) due to progressive quenching as the temperature increases during the heating [19]. (2) Another method is to record the glow curves at different heating rates [20]. When heating rate is increased, the TL glow peaks shift to higher temperatures. In case of thermal quenching the glow peak area decreases with increase of heating rate. The glow peak area, plotted as a function of glow peak temperature T_m, provides the quenching factor at different temperatures. These methods may not be useful in the case of high temperature glow peaks ($T_m > 350$ °C), which face interference from the thermal background and generally are quite weak in intensity, possibly due to the impact of strong thermal quenching. Kathuria and Sunta [21] envisaged that the thermal quenching factor of 400 °C glow peak of LiF:Mg, Ti (TLD-100) compared with the glow peaks appearing at temperatures below 300 °C, may be found by measuring the ratio between the decrease in the intensity of the 400 °C glow peak and the sum total of the PTTL intensity produced from room temperature to 300 °C by 254 nm UV exposure. The sample is exposed to a saturation dose ($\sim 10^3$ Gy) of gamma or beta rays and heated at 350 °C in order to retain the 400 °C peak as residual before exposure to UV light to induce PTTL. The UV exposures were of low fluence, so that no significant photo-bleaching takes place in the PTTL peaks. In order to achieve measurable decrease in the 400 °C residual glow peak, multiple low fluence UV exposures are given and the PTTL intensity measured after each UV exposure. The sum of all these PTTL intensity measurements provides the total PTTL produced. The 400 °C glow peak intensity is measured before starting the series of UV exposures and at the end of the last UV exposure. The difference between these two measurements provides a measure of the TL intensity transferred from the trap of the 400 °C glow peak to the traps of the PTTL glow peaks appearing up to 300 °C. The ratio of decrease in the intensity of the 400 °C glow peak and the summed up intensity of the PTTL peaks is the index of the thermal quenching at $\sim$400 °C relative to the temperature range of PTTL peaks namely room temperature to 300 °C. Assuming that 50 % of the untrapped carriers from traps of the 400 °C glow peak are captured in the traps of the PTTL glow peaks, and another 50 % may undergo recombination transition, it was estimated that the thermal quenching of TL at 400 °C may be about 85 times of that at temperatures below 300 °C. This value is obviously a rough estimate since there is likely to be a gradient in the quenching pattern of the temperature range below 300 °C also.

References

1. C.M. Sunta, Phys. Stat. Solidi. **37**, 81 K (1970)
2. C.M. Sunta, Phys. Stat. Solidi A **53**, 127 (1979)
3. C.M. Sunta, Radiat. Prot. Dosim. **8**, 25 (1984)
4. M.A. El-Kolaly, S.M.D. Rao, K.S.V. Nambi, A.K. Ganguly, Pramana **14**, 165 (1980)
5. C.M. Sunta, Proc. Third Int. Conf on Luminescence Dosimetry (now called Solid State Dosimetry) Danish AEC Riso Report No. 249, p. 392 (1971)
6. M.A. El-kolaly, Thermoluminescence studies of natural and doped CaF_2 phosphors, Bhabha Atomic Research Bombay, Ph.D. Thesis Bombay Univ., 1977
7. D. Curie, *Luminescence in Crystals* (Methuen and Co. Ltd., London, John wiley and Sons Inc., New York, 1963), p. 190
8. I.K. Bailiff, S.G.E. Bowman, S.F. Mobs, M.J. Aitken, J. Electrostat. **3**, 269 (1977)
9. C.M. Sunta, Radiat. Effects **79**, 149 (1983)

10. C.M. Sunta, S. Watanabe, J. Phys. D Appl. Phys. **9**, 1271 (1976)
11. W.B. Fowler (ed.), *Physcis of Color Centers* (Academic Press, New York and London, 1968)
12. C.M. Sunta, V.N. Bapat, S.P. Katuria, Proc. Third Int. Conf. Luminescence Dosimetry (in the Solid State Dosimetry (SSD) Series), Danish AEC Research Establishment Riso (Denmark) Report No. 249, p. 146 (1971)
13. E.S. Pasternack, A.M. Gaines, P.W. Levy, Bull. Amer. Phys. Soc. **22**, 409 (1977)
14. A.G. Wintle, D.J. Huntley, Nature **279**, 710 (1979)
15. A.G. Wintle, D.J. Huntley, J. Canadian, Earth Sci. **17**, 348 (1980)
16. P.W. Levy, PACT-6, J. European Study Group Physical, Chemical and Mathematical Techniques Applied to Archaeology, in *Proc. Specialist Seminar on TL Dating* (Oxford Res. Lab. Arch. and History of Art, 1982), p. 224
17. C.M. Sunta, B.C. Bhatt, P.S. Page, in *Proc. Third Int. Conf. on Luminescence and Its Applications (NCLA-2008)*, National Physical Laboratory New Delhi, ed. by S. Chawla, H. Chander, K.V.R. Murthy (McMillan India Ltd., 2008)
18. S.G. Gorbics, A.E. Nash, F.H. Atix, Proc. Second Int. Conf. Luminescence Dosimetry, Gatlinberg Tenn. USAEC Report Conf-680920 (1968), p. 568
19. C.M. Sunta, V.N. Bapat, PACT 6 (1982) p. 252
20. S.G. Gorbics, A.E. Nash, F.H. Atix, Proc. Second Int. Conf. Luminescence Dosimetry, Gatrinberg Tenn. USAEC Report Conf-680920, p. 587 (1968)
21. S.P. Kathuria, C.M. Sunta, Phys. Med. Biol. **26**, 707 (1981)

Appendix A
Symbols Used in the Book

Symbols	Meaning
a	M/N_i where M is the concentration of the thermally disconnected deep traps (TDDT) and N_i is the concentration of the interactive type of the active traps
A_h	Recombination coefficient of a free electron $A_h = \sigma_h v$ where σ_h is the cross section (cm^2) of the recombination centre for electron capture and v is the free electron velocity which is about 10^7 cm s^{-1}. The parameter A_h thus has units cm^3 s^{-1}
A_m	Capture coefficient of the thermally disconnected deep trap (TDDT) for free electron capture, $A_m = \sigma_m v$ where σ_m is the electron capture cross section for the TDDT and v = free electron velocity, which is about 10^7 cm s^{-1}. Thus units of A_m also are cm^3 s^{-1}
A_n	Capture (it means trapping or retrapping in the context of thermoluminescence) coefficient of a trap for free electron capture. $A_n = \sigma_n v$ where σ_n is the electron capture cross section for the trap and v is the free electron velocity
α	Alpha
b	Kinetic order
D	Dose, radiation dose
E	Energy, Activation energy
eV	Electron-volt, unit of energy
F	Recombining fraction. It is the fraction of the released (thermally) electrons from the active traps, which undergoes recombination to produce luminescence
Gy	Gray is used as unit of dose. It is the absorbed energy by an irradiated sample, 1 Gy = 1 J/kg
H	Total concentration of hole centres (recombination centres)
h	Concentration of occupied hole centres, it means the concentration of the available recombination centres at a given temperature T
I	Intensity, TL intensity. It is used also for total integrated intensity in a glow peak.
I_m	Maximum (peak) value of intensity in a glow peak. It may be used also for total integrated intensity
K	N_i/N_n
K	Absolute value of temperature measured in units of Kelvin

(continued)

C. M. Sunta, *Unraveling Thermoluminescence*,
Springer Series in Materials Science 202,
DOI: 10.1007/978-81-322-1940-8, © Springer India 2015

(continued)

Symbols	Meaning
M	Total concentration of thermally disconnected deep traps (TDDT) (cm^{-3} or m^{-3})
m	Concentration of filled TDDT
m_O	Concentration of filled TDDT at the beginning of heating to read the TL
N	Total concentration of the active traps
N_c	Effective density of states in the conduction band
N_i	Total concentration of interactive types of active traps
N_n	Total concentration of non-interactive type of active traps
n	Concentration of filled traps at any given temperature during the heating
n_O	Concentration of filled traps at the beginning of heating run
n_n	Concentration of non-interactive type of active traps
P	Probability of thermal excitation, $p = s\ \exp(-E/kT)$
s	Attempt to escape frequency or frequency factor (s^{-1})
T	Temperature
T_m	Temperature at the maximum intensity of the glow peak called also the peak temperature
T_R	Time taken in recording a glow peak, $T_R = (T_f - T_O)/\beta$ where T_f and T_O are the temperature at the end of the glow peak and at the start of heating and β is the heating rate
β	Heating rate ($K\ s^{-1}$)
μ	It is equal to δ/ω
v	Free electron velocity in the conduction band
σ	Cross section
σ_h	Electron capture cross section of recombination centres
σ_m	Electron capture cross section of TDDT
σ_n	Electron capture cross section of active traps
δ	Half width of the glow peak on the falling intensity side $\delta = T_2 - T_m$, where T_m is temperature at the maximum intensity of the glow peak and T_2 is the temperature at half of the TL intensity on the fall side of the glow peak
τ	Half width of the glow peak on the rising intensity side, $\tau = T_m - T_1$, where T_1 is the temperature at half of the peak intensity on the rising intensity side of the glow peak
ω	Full width at half the glow peak maximum $\omega = T_2 - T_1$ where T_1 and T_2 are the temperatures at half intensity of the peak maximum on the rising side of the peak and on its fall side
τ_{rec}	Recombination life time. It is the inverse of the recombination rate (radiative plus nonradiative). It represents the mean life of the free electron in the conduction band
τ_{max}	Maximum value of recombination life time

Appendix B
Acronyms of Technical Terms and Frequently Appearing Author Names

Acronym	Meaning
AT	Active traps. It means those traps which are getting thermally excited at a given temperature during recording of TL glow
DRM	Delocalized combination model
FO	First order (kinetics)
GO	General order (kinetics)
HCP	Heavy charged particles such as α particles and protons
IMTS	Interactive multi-trap system. The term means an energy level scheme which includes both the active traps and the thermally disconnected deep traps (TDDT). In this scheme the TDDTs recapture the electrons which have been ejected out from the active traps during the heating to read the TL
KO	Kinetic order
LED	Light emitting diodes
LET	Linear energy transfer-It signifies the energy transferred by an incident radiation particle per unit length of its path
LM	Lewandowski and McKeever
LRM	Localized recombination model
MO	Mixed order (kinetics)
MP	May and Partridge
NMTS	Non-interactive multi-trap systems. In this scheme also the deep traps exist, but they do not capture the free electrons ejected out from the active traps.
OTOR	one trap- one recombination centre
PPM	Parts per million
PRT	Parent residual trap
PTTL	Photo-transferred thermoluminescence
RC	Recombination centre
RT	Residual thermoluminescence
RW	Randall and Wilkins
SO	Second order (kinetics)
ST	Shallow traps
TL	Thermoluminescence
TLD	Thermoluminescence dose meter

C. M. Sunta, *Unraveling Thermoluminescence*,
Springer Series in Materials Science 202,
DOI: 10.1007/978-81-322-1940-8, © Springer India 2015

Index

C. M. Sunta, *Unraveling Thermoluminescence*,
Springer Series in Materials Science 202,
DOI: 10.1007/978-81-322-1940-8,